Handbook of Solid-state Troubleshooting

Handbook of Solid-state Troubleshooting

Hershal Gardner

RESTON PUBLISHING COMPANY, INC.
A Prentice-Hall Company
Reston, Virginia

Library of Congress Cataloging in Publication Data

Gardner, Hershal
 Handbook of solid-state troubleshooting.

 1. Semiconductors—Testing. 2. Electronic
apparatus and appliances—Maintenance and repair.
I. Title.
TK7871.85.H47 621.3815'2'028 76-832
ISBN 0-87909-339-0

© 1976 by Reston Publishing Company, Inc.
A Prentice-Hall Company
Reston, Virginia 22090

10 9 8 7 6 5

Printed in the United States of America

TECHNICAL DRAWINGS BY ROBERT MOSHER

Table of Contents

Preface

With the rapid advance of solid-state technology, the need has become apparent for a state-of-the-art semiconductor troubleshooting book. This textbook emphasizes the practical troubleshooting aspects of semiconductor devices, circuits, and systems, and holds theory to a minimum. A broad spectrum of topics has been chosen for treatment, including device testing, instruments, tools, component replacement, solid-state amplifier servicing, power-supply troubleshooting, control-circuit servicing, test-equipment maintenance and repair, radio receiver troubleshooting, FM stereo servicing, quadriphonic equipment troubleshooting, op-amp testing and repair, television receiver servicing, and digital equipment troubleshooting.

Semiconductor circuit analysis and testing is stressed throughout, in accordance with the prevailing state of the art. Topical treatment is instrument-oriented, inasmuch as efficient troubleshooting of present-day semiconductor circuitry demands the use of both basic and comparatively sophisticated test instruments. Accordingly, test techniques and procedures are explained that involve the VOM, TVM, signal tracer, oscilloscope, transistor checker, capacitor tester, stereo-multiplex generator, sweep generator, television analyzer, logic probe, logic clip, pulser, and digital word generator. The emphasis throughout the text is directed to symptom analysis, instrument application, and interpretation of test results.

It is assumed that the student has completed courses in basic electricity, electronics, radio theory, TV theory, and elementary digital-computer theory. It will also be helpful if the student has taken a

previous course in electrical and electronic instruments and measurements. However, an alert student can assimilate this text satisfactorily with the superficial knowledge of instruments gained from previous courses in physics and basic electricity and electronics. Mathematics has been held to a minimum, and graphical treatments of quantitative considerations have been utilized in this book. Nevertheless, it will be considerably easier for the student to understand the reasoning underlying troubleshooting procedures and data evaluation if he has completed courses in algebra, geometry, and elementary trigonometry. The author is indebted to numerous manufacturers, as credited throughout the text, for illustrative material and technical data.

This book is the outcome of many years of teaching experience, both on the part of the author and of his associates. In a significant sense, this text represents a team effort, and the author gratefully acknowledges the constructive criticisms and contributions of the City College teaching staff. It is appropriate that this work be dedicated as a teaching tool to the instructors and students of our technical schools and junior colleges.

HERSHAL GARDNER

Handbook of Solid-state Troubleshooting

Solid-state Principles and Testing Procedures

1.1 ELEMENTS OF SEMICONDUCTOR TECHNOLOGY

There are numerous types of semiconductors. A semiconductor has a resistivity value between that of metals and that of insulators, in which the number of electrical-charge carriers increases with increasing temperature over some range. As an illustration, the number of electrons and holes in a conventional germanium diode increases as the temperature of the device rises above room temperature. Next, if the temperature of the device is increased excessively, the diode will be destroyed. Note that metals such as copper or iron have an opposite temperature characteristic. In other words, the resistance of a copper wire increases as its temperature is increased. In turn, copper has a positive temperature coefficient, whereas a germanium diode has a negative temperature coefficient. Of course, a copper wire will also be destroyed if its temperature is increased excessively.

Effective troubleshooting of semiconductor devices, circuits, and systems requires a basic understanding of device action, circuit action, and system operation. Figure 1–1 illustrates the basic types of semiconductor devices: photoconductors, thermistors, signal diodes, rectifiers, zener diodes, compensating diodes, varactors, bipolar transistors, unipolar transistors, thyristors, silicon controlled rectifiers, triacs, and integrated circuits. An example of semiconductor resistance variation is depicted in Fig. 1–2 for a cadmium-sulfide photoconductive

Figure 1–1. Typical semiconductors including transistors, diodes, unijunction transistors, high-power rectifiers, SCR's and triacs, with current capabilities from 1mA to 2,000A, and voltage ratings up to 3,000V. *(Courtesy of General Electric Semiconductor)*

cell, or light-dependent resistor (LDR). Its resistance varies from 100 ohms in bright light to 0.5 megohm in complete darkness. We shall find that the resistance of an LDR is the same, regardless of ohmmeter polarity. In other words, if the test leads of the ohmmeter are reversed, the resistance reading remains the same.

Thermistors provide another useful example of semiconductor resistance variation. A thermistor is basically a bead of certain metallic oxides. As its temperature is increased from 0° to 100°, for example, the resistance of the device decreases typically from 1000 ohms to

100 ohms. If the temperature of the thermistor is further increased to the point of burnout, its resistance decreases typically to a value of 10 ohms. We shall find that the resistance of a thermistor is the same, regardless of ohmmeter polarity. Any thermistor will be damaged if it is operated above its rated temperature limit. On the other hand, an LDR will not be damaged by exposure to extremely bright light, although it will be destroyed if subjected to excessively high temperature.

A semiconductor diode has a forward-resistance (low resistance) value and a reverse-resistance (high resistance) value, as depicted in Fig. 1–3. As an illustration, if a certain ohmmeter is operated on its R × 10 range, we may find that a germanium diode shows a forward resistance of 70 ohms and a reverse resistance of infinity. Then, if another ohmmeter is employed that applies a somewhat different value of test voltage to the device, we may read a forward resistance of 100 ohms. However, the essential point here is that a normal operating semiconductor diode has a low value of forward resistance and a very high value of reverse resistance. Another basic point concerns the voltage drops across forward-biased junctions. Thus, under typical current flow, a germanium diode will show approximately 0.25-volt drop across its forward-biased junction. On the other hand, a silicon diode will show approximately 0.6-volt drop across its forward-biased junction. These topics are considered in greater detail subsequently.

Bipolar transistors have two junctions, and each junction has

Figure 1–2. Resistance measurement of a CdS light-dependent resistor.

Figure 1–3. Measurement of forward and reverse resistance values. **(a)** Forward resistance is 70 ohms; **(b)** Reverse resistance is infinite.

a forward-resistance value and a reverse-resistance value. Usually, the emitter-base forward-resistance value is somewhat smaller than the collector-base resistance value, as exemplified in Fig. 1–4. A few bipolar transistors will show the same value of forward resistance for the emitter-base and collector-base junctions. If the two junctions have unequal forward-resistance values, then the resistance from emitter to collector will depend upon ohmmeter polarity. This is always a high resistance value, and the ohmmeter must be operated on a high range, such as R × 1000. In the example of Fig. 1–4, the emitter-collector forward resistance has a value of 1 megohm. Note that if the transistor were a PNP type, all of the test polarities noted would be reversed.

Junction field-effect transistors show a different pattern of resistance readings, as depicted in Fig. 1–5. In other words, forward and reverse readings occur between gate and source, or between gate and drain only; the resistance reading from source to drain is the same, regardless of ohmmeter polarity. Note in passing that the resistance measured from source to drain is also the same, regardless of the ohmmeter range that is used. In other words, the source-drain resistance is linear. On the other hand, the junction resistance from gate to source, or from gate to drain, is nonlinear. The practical importance of this nonlinearity is that the forward-resistance value that is meas-

Figure 1–4. Example of bipolar transistor junction resistance values.

Figure 1–5. Example of junction-FET resistance values.

ured across the junction will be different if the ohmmeter range is changed. Note also that ohmmeter polarity is not necessarily indicated correctly by the test-lead color coding. In other words, the red lead may be negative and the black lead positive. Therefore, it is good practice to check ohmmeter polarity with a dc voltmeter, as shown in Fig. 1–6.

Insulated-gate field-effect transistors (IGFET's) are also called metallic-oxide substrate field-effect transistors (MOSFET's). They are different from junction field-effect transistors (JFET's) in that there is no junction between the gate and the semiconductor source-drain substance. In turn, a MOSFET shows infinite resistance between the gate and the source, or between the gate and the drain. However, a MOSFET shows a definite resistance value between source and drain terminals, in the same manner as a JFET. Note carefully that an ordinary MOSFET is very likely to be damaged if its gate terminal is disconnected from its source or drain terminal. This danger is present because of the extremely thin insulation provided under the gate electrode. A "floating" gate is likely to pick up enough static electricity to break down the insulation and ruin the MOSFET. Therefore, be sure to provide some kind of metallic connection from the gate terminal to the source or drain terminal whenever the MOSFET is removed from its circuit.

Dual-gate MOSFET's are basically the same as ordinary MOSFETS, except that two gate electrodes are provided. They are used in comparatively elaborate circuits. Insofar as handling is concerned during troubleshooting procedures, the same precaution applies

Figure 1–6. Voltmeter pointer moves upscale if the red ohmmeter lead is a positive lead.

Figure 1–7. Internal arrangement of a dual-gate protected MOSFET.

as noted above. In other words, it is important to provide a short-circuit between each gate terminal and the source terminal whenever the dual-gate MOSFET is not connected into its circuit—neither of the gate electrodes should ever be permitted to "float." However, this precaution does not apply to dual-gate protected MOSFET's, such as depicted in Fig. 1–7. In other words, this type of MOSFET contains built-in back-to-back zener diodes that bypass any excessive gate voltage to the source electrode. In turn, a dual-gate protected MOSFET can be handled without any precautions against static electricity.

When the gate-to-source resistance of a gate-protected MOSFET is checked with an ohmmeter, the resistance reading depends greatly upon the test voltage applied by the meter. Many ohmmeters do not apply sufficient test voltage to cause zener conduction, and such instruments will show an infinite-resistance reading for the transistor if it is in normal condition. On the other hand, an occasional ohmmeter, particularly when operated on its highest range, will apply sufficient test voltage to cause zener conduction. In such a case, the instrument will show a very low resistance reading. Considerations of semiconductor resistance measurements are explained in greater detail subsequently.

1 . 2 BASIC SEMICONDUCTOR DEVICE TESTERS

Service-type semiconductor device testers are typically designed to check bipolar transistors, FET's, diodes, silicon-controlled rectifiers (SCR's), triacs, and unijunction transistors. Note in passing that an SCR is a three-junction semiconductor device that is normally an open circuit until an appropriate gate pulse is applied to the gate terminal, at which time it rapidly switches to the conducting state. Its action

is comparable to that of a thyratron electron tube. SCR circuitry and operation is discussed in greater detail subsequently. Triac is a trade name (General Electric) for a gate-controlled full-wave ac silicon switch designed to switch from a blocking state to a conducting state for either polarity of applied voltage with positive or negative gate triggering. A unijunction transistor (UJT) is also called a silicon double-base diode or silicon unijunction transistor. It is a three-terminal semiconductor device exhibiting stable open-circuit, negative-resistance characteristics. These topics are treated in greater detail in a following chapter.

Other service-type semiconductor testers have more limited capabilities and can check only bipolar transistors and diodes. All semiconductor testers can be classified as out-of-circuit instruments, in-circuit instruments, or combination in-and-out-circuit instruments. In-circuit instruments are basically of the go/no-go type; in other words, a low-impedance testing arrangement is employed that merely indicates whether the transistor provides any gain. It is impractical to measure transistor gain with an in-circuit test, because the transistor is necessarily shunted by unknown circuit impedances. For this reason also, it is impossible to determine whether a transistor has junction leakage with an in-circuit test. In-circuit diode tests are limited to detecting open-circuited or "dead-short" diodes. Because of unknown circuit impedance, it is impossible to measure the front-to-back ratio of a diode with an in-circuit test.

In spite of the inherent limitations of an in-circuit transistor tester, this type of instrument nevertheless provides very useful troubleshooting data. It shows whether a transistor is workable or unworkable in most situations. In turn, if a transistor is found to be workable with an in-circuit test, but the stage shows a serious mal-function, it is logical to conclude that the trouble will be found in a circuit component instead of the transistor. When the results of an in-circuit test are inconclusive, the transistor can be disconnected from its circuit for an out-of-circuit test. The results of an out-of-circuit test are always completely conclusive. As a practical note, remember that if the technician does not wish to unsolder a transistor from its circuit board for test, two of the transistor PC leads can be "cut" for temporary disconnection. After the out-of-circuit test is completed, the PC lead continuity can be restored by small drops of solder across the cuts.

An out-of-circuit semiconductor tester typically measures bipolar transistor current gain in terms of beta values. The beta value is defined as the ratio of collector current to base current, with the

transistor operating in the CE mode. A typical small-signal bipolar transistor normally has a beta value on the order of 100. Note that particular types of transistors might be rated for a beta value as low as 50, or as high as 300. Power-type bipolar transistors are generally rated for somewhat lower beta values than are small-signal transistors. Out-of-circuit semiconductor testers also provide measurement of leakage current. A small-signal bipolar transistor normally has very low junction leakage (reverse current), which is measurable in micro-amps, or picoamps for silicon. On the other hand, the normal collector current is measured in milliamperes. Power-type bipolar transistors tend to have at least a slight amount of junction leakage in normal operation.

Out-of-circuit semiconductor testers designed for checking FET's provide a transconductance function. In other words, an FET cannot be tested for gain in terms of beta values. This impossibility follows from the fact that the gate of an FET does not draw current, at least in theory. Therefore, a current-ratio value is meaningless. A transconductance test of an FET provides a measure in micro "siemens"* of the ratio of gate voltage to drain current for a small-signal condition. Diodes are checked out-of-circuit for forward-current and reverse-current values. Note that these values may not be entirely consistent from one instrument to another, because the amount of forward current depends considerably on the applied test voltage. On the other hand, the reverse current in a normal diode is essentially independent of the applied test voltage; as noted previously, this reverse current is normally extremely small.

A typical in-circuit transistor tester is illustrated in Fig. 1–8. It provides a good-bad in-circuit gain test, and an out-of-circuit leakage test. The good-bad test checks a bipolar or field-effect transistor in the common-emitter or common-source configuration, and indicates whether the device will provide an inverted square-wave output in response to an applied square-wave input. If the device develops an inverted square-wave output, an audible tone is produced by the instrument. Otherwise, there is no sound output; lack of a sound output indicates that the transistor is unworkable, or "bad." A practical feature of this in-circuit test is that the three test leads from the instrument can be connected in any order whatsoever to the terminals of the transistor under test. In other words, the instrument contains basic computer-logic circuitry and a switching arrangement for automatically making correct test connections.

*A siemen is equal to a mho.

Figure 1–8. A typical in-circuit transistor tester. *(Courtesy of Sencore)*

With reference to Fig. 1–8, six pushbuttons and a polarity switch (NPN-PNP) are provided to accommodate both basic types of bipolar transistors and to determine the emitter, base, and collector terminals. This operating facility also accommodates FET's. To test a transistor in circuit, the technician connects the three test leads in any order to the transistor terminals, and presses the pushbuttons in sequence. If no sound output is obtained for any pushbutton, the technician reverses the setting of the polarity switch and again presses the pushbuttons in sequence. An audible "chirp" from the instrument for any pushbutton test indicates a "good" transistor; silence indicates that the transistor is "bad." There are rare instances in which a "good" transistor will test "bad" in circuit; for example, if there is a resistance of less than 100 ohms from base to collector, or a capacitance of more than 50 μf from collector to base, the in-circuit test will indicate that the transistor is "bad." To avoid the possibility of doubt in such a situation, the technician may cross-check by means of an out-of-circuit test after the device has been disconnected from its circuit board. Then, if the gain indication is "good," and the transistor does not have excessive leakage, it is concluded that the trouble symptom

is being caused by a circuit component and not by the transistor. Table 1–1 tabulates the maximum tolerable leakage currents for various common types of transistors.

With reference to Table 1–1, recall that I_{CBO} is the leakage current that flows in a transistor when a test voltage is applied between the collector and base, with the emitter open and the collector-base junction reverse-biased. In an FET, this is called I_{GDO}. Next, I_{BEO} is the current that flows through the forward-biased base-emitter junction. In an FET, this is called I_{SGO}. Again, I_{CEO} is the leakage current that flows in a transistor when the test voltage is applied between emitter and collector with the base open, and the emitter-collector junction reverse-biased. In an FET, this is called I_{DSO}. Again, I_{BEO} is the leakage current that flows through the reverse-biased emitter-base junction with the collector open. In an FET, this is called I_{GSO}. Again, I_{CEO} is the leakage current that flows through a transistor when the test voltage is applied between collector and emitter, with the base open. In an FET, this is called I_{DSO}. Finally, I_{BCO} is the current that flows through a reversed-biased base-collector junction. In an FET, this is called I_{DGO}.

The transistor tester illustrated in Fig. 1–8 contains a regulated power supply with $+5$-volt and -5-volt outputs. This power supply energizes the 2-kHz oscillator and signal-amplifier arrangement de-

TABLE 1-1

Maximum Tolerable Leakage Currents for Various Transistors

TYPE OF DEVICE	BUTTON 1 I_{CBO} (GDO)	BUTTON 2 I_{BEO} (SGO)	BUTTON 3 I_{ECO} (DSO)
Small Si. Transistor Large Si. & Small	0-.1μA	3000μA	0-.1μA
Ge. Transistor	1-50μA	3000μA	1-50μA
Large Ge. Transistor	10-3000μA	3000μA	10-3000μA
JFET	0μA	3000μA	3000μA
MOSFET	0μA	0μA	3000μA

	BUTTON 4 I_{EBO} (GSO)	BUTTON 5 I_{CEO} (SDO)	BUTTON 6 I_{BCO} (DGO)
Small Si. Transistor Large Si. & Small	0-.1μA	0-1μA	3000μA
Ge. Transistor	1-50μA	10-500μA	3000μA
Large Ge. Transistor	10-3000μA	100-3000μA	3000μA
JFET	0μA	3000μA	3000μA
MOSFET	0μA	3000μA	0μA

(a)

(b)

Figure 1–9. Oscillator circuit section of transistor tester. **(a)** Regulated power supply; **(b)** Square-wave generator and amplifier.

picted in Fig. 1–9. It includes an integrated-circuit oscillator and an amplifier comprised of the push-pull emitter-follower transistors TR1 and TR2. A square-wave signal with an amplitude of 3 volts p-p is developed across R15. Next, with reference to Fig. 1–10, the transistor under test is driven in the CE mode, with its collector connected to R19. If the transistor is good, an amplified out-of-phase signal flows back through R19 and is applied to C6. Note in Fig. 1–10(a) that the base and collector signals are 180 deg out of phase.

If the reader is unfamiliar with computer-logic circuitry, it will

Figure 1–10. Logic circuitry and indicator drive of transistor tester. **(a)** Simplified test circuit; **(b)** Detailed circuit.

be advisable for him to omit the remainder of this topic, and to proceed instead to Topic 1.3. However, readers who are reasonably familiar with computer logic can read the remainder of this topic to advantage. Note in Fig. 1–10 that the base-driving signal from the 2-kHz oscillator is amplified by integrated circuit 1C (IC1C), and that the collector output signal is amplified by IC1D. In turn, the outputs from IC1C and IC1D are coupled to an exclusive "OR" gate that provides an output signal to the speaker and meter only when the two input signals are 180 deg out of phase. Note that R20 is included in the instrument circuitry to insure that when the test leads are open, or when a defective transistor is connected, the signal at the junction of R18 and R19 will be in phase with the base-driving signal.

The 2-kHz test signal is coupled to the noninverting input of IC1D in Fig. 1–10 by R20, C6, and R22. Since the resistance of R20 is large in comparison with R18 and R19, the inverted signal from the collector of a good transistor under test will override the in-phase signal from R20. Note that the logic circuitry compares the base and collector signals from the transistor under test and provides the output to the meter and speaker. IC2 is a quad, two-input "NAND" gate wired as an exclusive "OR" gate. The only time that the output from the gate is zero occurs when both inputs are positive. With a good transistor under test, the base signal coupled into input 2 of IC2A will be out of phase with the collector signal coupled into input 1. Since both inputs of IC2A are never positive at the same time, the ouput at pin 3 remains positive. This positive output from pin 3 is coupled into input pin 5 of IC2B and input pin 9 of IC2C. IC2B and IC2C then function as inverters, coupling the inverted base and collector signals into the inputs of IC2D.

Since both inputs of IC2D are never positive at the same time, the output pin 11 remains positive. This positive output is coupled to M1, and deflects the pointer to the "good" sector on the scale. Also, the square-wave signal at output 8 of IC2C is amplified by TR3 to drive the speaker for an audible output indication. With a "bad" transistor under test, the inputs of IC2A will be in phase, resulting in a square-wave output at pin 3 that is 180 deg out of phase with the input signal. This out-of-phase square wave is coupled to the inputs of IC2B and IC2C along with the in-phase signals from base and collector. Since both inputs of IC2B and IC2C are never positive at the same time, the outputs of IC2B and IC2C are always positive. The positive outputs IC1B and IC1C are coupled to the inputs of IC2D, resulting in a zero output from IC1D. This zero output is coupled to meter M1, with the result that a "bad" scale indication is

obtained. Also, output 8 of IC2C is always positive, and no signal is coupled into the audio transistor and speaker, with the result that no sound output occurs.

Some technicians prefer to use a curve tracer for testing solid-state devices. Figure 1–11 shows the circuit of the RCA Quicktracer, and the out-of-circuit hookup arrangement. This tester also has considerable in-circuit utility. The illustrations in Figure 1–12 show the waveforms of the Quicktracer in "open" and "shorted" positions. Thus, when the test leads are applied to a circuit component, the waveform on the scope screen will change depending on the resistance and/or reactance of the component. The Quicktracer is set up as follows:

1. Turn the oscilloscope on, and set the horizontal-input selector to "external."
2. Adjust the scope controls to display a dot at the center of the screen.
3. Connect the horizontal, ground, and vertical outputs of the Quicktracer to the corresponding scope terminals, and plug in the Quicktracer.
4. Adjust the scope horizontal gain control to provide adequate horizontal deflection.
5. Short-circuit the test leads of the Quicktracer together and adjust the scope vertical gain control for adequate vertical deflection.

Consider the patterns obtained when a diode is under test. We proceed by connecting the test leads to the diode and observing the resulting waveform display. If the trace deflects downward to the right, the cathode is connected to the black test lead. If there is no vertical deflection, the diode is "open." On the other hand, a straight vertical line indicates a "shorted" device, except when one is testing in power-supply circuits wherein the shunt current through the filter capacitors causes an apparent "short." If leakage is indicated by a "slanted" horizontal deflection, it can be determined whether this is a diode fault or a shunt-circuit effect by disconnecting one end of the diode, or by comparison of the displayed waveform with a standard waveform. Figure 1–13 shows typical waveforms for a normal diode. Figure 1–14 exemplifies waveforms produced when zener diodes are tested. Downward deflection at the right is caused by normal diode conduction. Upward deflection at the left occurs when the applied voltage exceeds the zener voltage of the diode.

Next, consider the testing of bipolar transistors. A bipolar

a - Circuit of the quicktracer.

b - Out-of-circuit hookup arrangement.

Figure 1–11. Circuit of the Quicktracer, and hookup arrangement for a typical out-of-circuit component test. *(Courtesy of RCA)*

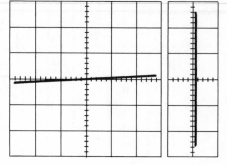

Figure 1-12. A horizontal trace is displayed when there is an open-circuit between the test leads; a short-circuit displays a vertical trace. The scope may be adjusted to produce the desired lengths. *(Redrawn by permission of RCA)*

transistor may be regarded as two diodes connected in series with opposing polarities. If the transistor is a PNP type, the base material serves as a cathode; the emitter and collector are the two anodes. In turn, if the black lead from the Quicktracer is connected to the base of the transistor, connecting the other test lead to either emitter or collector will produce downward deflections at the right end of the trace. On the other hand, if an NPN transistor is tested, deflection will be upward at the left, because the base serves as an anode, rather than a cathode. Figure 1–15 shows typical displays produced by good transistors.

The terminals of most transistors and their type (PNP or NPN) can be determined by the following procedure:

1. Connect the test leads at random to the transistor terminals. When the waveform displayed approximates any of the ones specified below, the leads are connected across the base-emitter junction.
2. Connect the red test lead to the collector.
3. Connect the black test lead alternately to the two remaining terminals. The vertical deflection will be greater when the black test lead is connected to the base of the transistor.
4. When the black test lead is connected to the base, the vertical deflection will be downward at the right and upward at the left for most NPN transistors.

Figure 1–13. Waveforms generated when a diode is connected to the test leads. The one at the left results when the black test lead is connected to the diode cathode. Reversing the diode connections produces the waveform at the right. *(Courtesy of RCA)*

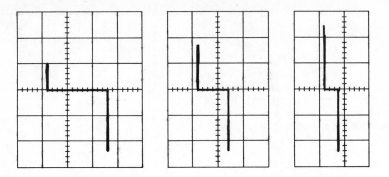

Figure 1–14. Waveforms produced by Zener diodes. The length of the horizontal section indicates Zener voltage. From left to right, 30, 15, and 7-volt Zeners were tested. Scope horizontal sensitivity in this series of waveforms is about 12.5 volts/cm. *(Redrawn by permission of RCA)*

Most transistors show a zener-type conduction through their base-emitter junction. However, for those transistors that do not show zener conduction, such as germanium types, the first step cannot be carried out, and the following procedure should be followed instead:

1. Connect the test leads at random to the transistor terminals. When an open circuit is indicated, the leads are connected to the collector and emitter. (A little conduction may occur near the voltage crossover point).
2. Connect the black test lead to the base of the transistor.
3. With the black test lead connected to the base, vertical deflection will be downward at the right for PNP transistors, and upward at the left for NPN transistors.

Typical normal waveforms for PNP and NPN transistors are exemplified in Fig. 1–15. Transistor leakage will be indicated by angular displacement of the horizontal segment of the waveform on the scope screen. Any leakage indicated in a silicon transistor should be considered a defect. However, in some germanium transistors, a little tilt of the horizontal segment may be permissible. An angle of 10 deg with respect to horizontal indicates approximately 5700 ohms of leakage resistance.

Next, consider the essentials of in-circuit transistor testing. In the absence of reference data, the foregoing techniques may be utilized to test transistors and identify their terminals either in circuit or out of circuit. However, the Quicktracer cannot distinguish between transistor leakage resistance and shunt resistance of the

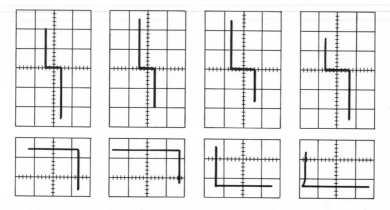

Figure 1-15. Typical PNP and NPN base-emitter waveforms are shown in (a) and (c). Typical PNP and NPN waveforms with red lead connected to collector are shown in (b) and (d). Greater vertical deflection results when the black test lead is connected to the base (lower left waveform of each set). The lower right waveform in each set results when the black lead of the Quicktracer is connected to the emitter. *(Redrawn by permission of RCA)*

circuit associated with the transistor. The tilt of the displayed waveform is a function of the total resistance across a junction. The correlation between the angle of displacement in the horizontal portion of the waveform and shunt resistance across the junction is indicated approximately by the waveforms shown in the resistance measuring procedure, as explained below. If the vertical section of the waveform is tilted, the transistor is faulty, since no external circuit can increase the zero resistance of a forward-biased junction.

If reference data are available, a comparison of the displayed waveform with these data is the most reliable testing technique. If the transistor terminals cannot be identified by inspection, proceed with the tests described above in order to make this identification. If

Figure 1-16. The "slope" of the trace indicates the resistance between the test leads of the Quicktracer. A short-circuit produces a vertical trace and an open-circuit produces a horizontal trace. The resistances used to produce the waveforms (left and down): were 100, 220, and 470 ohms. (Right and down): 1000, 2.2K, 4.7K, 6.8K, and 10K ohms. *(Redrawn by permission of RCA)*

the waveforms displayed during tests are not very nearly the same as the examples shown in Fig. 1–16, a defect may be presumed to exist. An open-circuit indication (straight horizontal line) always corresponds to an open junction. Various other abnormal waveforms can result from either a defect in a transistor, or the characteristics of the associated circuitry.

1 . 3 TOOLS FOR SEMICONDUCTOR TROUBLESHOOTING

Most semiconductor devices are soldered into printed-circuit boards. In turn, soldering and desoldering tools are basic requirements. A pencil-type soldering iron such as shown in Fig. 1–17(a) is used to solder replacement-part connections. A desoldering tool, depicted in Fig. 1–17(b), provides considerable convenience in disconnecting defective semiconductors or other components from a PC board. A heat-sink tool, shown in Fig. 1–18(a), is essential to prevent heat damage to semiconductor devices during soldering operations. It is a small clamp that is placed on a semiconductor lead between the point of soldering and the case of the device. An anti-wicking tool, seen in Fig. 1–18(b), is useful to prevent the flow of molten solder under the insulation of a wire. A snipper, such as that shown in Fig. 1–18(c), is useful for cutting insulation from the ends of wires.

A desoldering tool must be used properly, or difficulties will be encountered. Allow the desoldering tool to heat completely, before it is applied to the circuit board. Approximately 4 minutes is sufficient. Then, depress the bulb on the desoldering tool, and apply the hot tip to the joint that is to be desoldered. When the solder starts to flow, release the bulb, and the molten solder will be removed by vacuum action. It is good practice to immediately discharge the waste solder after each desoldering operation. To do so, place the tip of the tool over a suitable waste receptacle and squeeze the bulb rapidly a few times. Note carefully that the waste solder should never be reused. In other words, this "dead" solder is seriously oxidized, and if it is reused, the joint will be defective both electrically and mechanically. This type of joint is called a "cold-solder" connection.

It is also good practice to keep the tip of the desoldering tool clean by frequently wiping the tip on a damp sponge or lint-free cloth. The tip orifice can be cleaned while the tool is hot by inserting a wire into the opening of the tip. Note that no attempt should be made to remove a tip while the tool is cold. After the tool is com-

pletely heated, its tip can be removed without danger of damage. The useful life of the tool and tip can be increased if the tip is kept tinned, and the waste solder is discharged after each desoldering operation.

Basic tools for disassembly and reassembly of semiconductor

(a)

(b)

Figure 1–17. Basic soldering equipment for semiconductor troubleshooting. **(a)** Pencil-type soldering iron. *(Courtesy of Utica Tool Co.)*; **(b)** Desoldering tool. *(Courtesy of Weller Tools)*

(a)

(b)

(c)

Figure 1–18. Supplementary tools for soldering operations. **(a)** Heat sink; **(b)** Anti-wicking tool; **(c)** Snipper.

equipment include an assortment of large and small screwdrivers, as illustrated in Fig. 1–19. Phillips-head screwdrivers (Fig. 1–20) are also required. Nut drivers, shown in Fig. 1–21, are a practical necessity for loosening and tightening hex nuts of various sizes. Larger nuts require the use of box end wrenches and adjustable wrenches, as ex-

emplified in Fig. 1–22 and 1–23. Combination pliers and long-nose pliers are necessities, and may be usefully supplemented by electrician's pliers, seen in Fig. 1–24. Note that all-purpose electrician's pliers are convenient for cutting machine screws to precise lengths without damage to the threads. This tool also serves as a wire cutter, stripper, and crimping tool. In addition, slim-nose, curved chain-nose,

Figure 1–19. A set of standard screwdrivers. *(Courtesy of Xcelite)*

Figure 1–20. Phillips-head screwdrivers. *(Courtesy of Xcelite)*

Figure 1–21. Assortment of hex nut drivers. *(Courtesy of Xcelite)*

Figure 1–22. A set of combination box end wrenches. *(Courtesy of Xcelite)*

Figure 1–23. Typical adjustable end wrenches. *(Courtesy of Xcelite)*

Figure 1–24. Typical troubleshooter's pliers. *(Courtesy of Starnetics)*

duck-bill and compound pliers are frequently used (Fig. 1–25). Hex and spline key wrenches, such as those shown in Fig. 1–26, will be needed occasionally.

Large and small tweezers are often helpful to retrieve nuts or other small objects from miniaturized semiconductor equipment. An ordinary dental mirror (Fig. 1–27) is sometimes a great convenience

Figure 1–25. Assortment of slim-nose, curved chain-nose, duck-bill, and compound pliers. *(Courtesy of Starnetics)*

when it is necessary to inspect behind or under a circuit-board component. A magnifying glass is occasionally needed when one is inspecting a PC board for microscopic cracks that can cause intermittent operation. Good lighting is essential in this situation, and a small high-intensity lamp should be available. Cleaning cloths, an assortment of soft brushes, a small blower, and standard cleaning solutions will be needed. Aerosol solvent cleaner is helpful in removing corrosion, oxide buildup, or other foreign substances. Spray contact cleaner in a container with an extension nozzle is used to troubleshoot intermittent switches.

Figure 1–26. Typical hex and spline key wrenches. *(Courtesy of Starnetics)*

Figure 1–27. Dental inspection mirror and tweezer assortment. *(Courtesy of Claude Michel)*

1 . 4 MOUNTING OF SEMICONDUCTOR DEVICES

Semiconductor devices are usually (but not always) mounted on PC boards. The devices may be plugged into sockets, although it is common practice to solder device leads to circuit-board conductors. A typical PC board without any components mounted in place is depicted in Fig. 1–28. When semiconductor devices and associated components are mounted on a board, the assembly has an appearance such as illustrated in Fig. 1–29. This is an example of modular construction. In other words, the circuit-board terminals are all located at one end of the board, and are designed for insertion or withdrawal from the main chassis. Modular construction facilitates troubleshooting procedures, since a suspected circuit board can be quickly checked by replacement. If the replacement board restores normal operation, the defective board can be set aside and repaired at leisure.

Power transistors are secured to a heat sink or to a metal chassis with a locknut, as shown in Fig. 1–30. Unless a power transistor makes firm contact with its heat sink, it will overheat and become defective. A film of silicone grease may be applied between the transistor case and its heat sink to insure good thermal conduction. When one is replacing a transistor, it is important to connect the leads correctly. Various basing arrangements are in general use, and Fig.

Figure 1–28. A typical PC board, minus devices and components.

(a) (b)

Figure 1–29. Circuit board module with devices and associated components.
(a) Bottom view; **(b)** Top view.

Figure 1–30. Mounting of a power transistor.

E=Emitter
B=Base
C=Collector

(Bottom views)

A=Anode
C=Cathode
G=Gate

Figure 1–31. Basing identifications for common small-signal and power-type transistors.

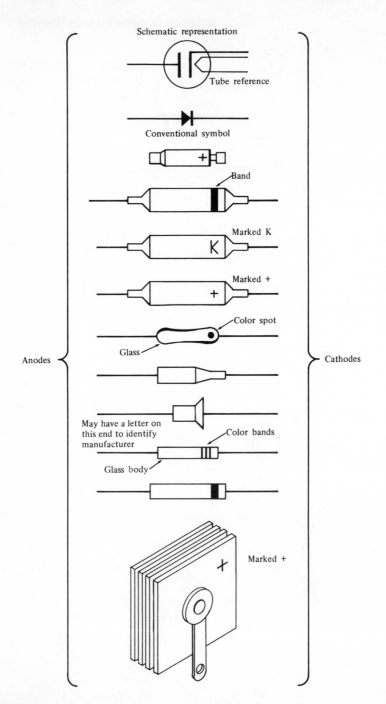

Figure 1–32. Various methods of indicating diode polarity. *(Redrawn by permission of Heath Company)*

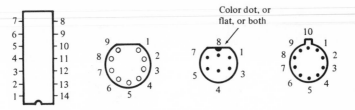

Figure 1–33. Basing identifications for common integrated circuits.

1–31 shows the more usual basing identifications for both small-signal and power-type transistors. Similarly, when one is replacing a small-signal diode or a power diode, it is essential to observe correct polarity. Figure 1–32 shows the various methods that are in common use to indicate diode polarity. Integrated circuits have more terminals than other semiconductor devices; Fig. 1–33 depicts the more usual basing identifications for integrated circuits.

1.5 CIRCUIT-BOARD SOLDERING PROCEDURES

Correct circuit-board soldering procedure is depicted in Fig. 1–34. After the replacement part is in position with its leads inserted through the terminal holes, it is good practice to bend the leads outward to hold the component in place during the soldering operation. Then apply the tip of the soldering iron so that it firmly contacts both the circuit-board foil and the wire or lug to be soldered, as shown in Fig. 1–34. The iron should be held so that solder is not likely to flow to adjacent foil conductors or connections. The solder should immediately be placed between the iron and the joint to be soldered. Remove the length of solder as soon as its end begins to melt and flow onto the lead and foil. Hold the tip of the iron in place only until the solder begins to flow outward over the foil. Then remove the iron quickly.

A soldering pencil or a small iron (approximately 25 watts) is recommended for circuit-board work. If a soldering gun or high-wattage iron is used, precautions must be taken to avoid circuit-board damage owing to overheating and excessive solder flow. The use of an excessive amount of solder increases the possibility of bridging between foil conductors or plugging holes that should be left open. If solder accidentally bridges across insulating areas between conductors, it can be cleaned off by heating the connection carefully and quickly wiping or brushing the solder away with a soft cloth or clean brush.

Figure 1–34. Correct soldering procedure, and common errors.

Holes that become plugged can be cleared by heating the area immediately over the hole while gently pushing the lead of a resistor, for example, through the hole from the opposite side, and withdrawing the lead before the solder hardens. Do not force the lead through; too much pressure applied before the solder has time to harden may separate the foil from the board.

Another method of parts replacement consists in clipping the defective component from the circuit board with a pair of diagonal cutters, leaving approximately ⅛ in. of lead projecting above the board. In the case of a composition resistor, it is sometimes helpful to crush the body of the resistor with a pair of pliers, and then to cut the remaining leads to suitable length. Then, the leads of the replacement component are cut so that the component will be positioned properly when the ends of the new leads are mated with the ends of the old leads on the circuit board. A small spiral connector is generally used to hold the ends of the leads mated correctly while soldering; the connector also adds mechanical strength to the arrangement.

REVIEW QUESTIONS

1. What is the purpose of a thermistor?
2. How is a heat sink used while soldering semiconductor connections?
3. What is the purpose of an anti-wicking tool?
4. Explain the uses of electrician's all-purpose pliers.
5. What is the primary advantage of modular construction?
6. Why is a coating of silicone grease applied between a power transistor and the heat sink?
7. Draw several diodes and indicate the various methods of indicating polarity.
8. State two reasons why extreme care must be taken when soldering semiconductors on printed circuit boards.
9. What is the advantage of clipping the leads of a defective part and soldering a new part to the old leads?
10. What is the purpose of the small spiral connector when using the technique mentioned in question 9?
11. Explain the basic distinctions between bipolar transistors and junction field-effect transistors.
12. How does a MOSFET differ from a JFET?
13. Describe a method of protecting a MOSFET from damage by static electricity.
14. Discuss the basic characteristics of an SCR.
15. State the two general types of semiconductor testers, and explain their differences in application.
16. Explain the chief operating features of a curve tracer for testing semiconductors.
17. Define collector-junction leakage in a transistor.
18. Why should solder collected by a desoldering tool be discarded?
19. What is the meaning of the term "modular construction"?
20. Describe the action of a heat sink utilized with a power transistor.
21. Name several ways of indicating diode polarity.
22. How many terminals will be found on typical integrated circuits?
23. Explain how excess solder should be removed from a PC board.
24. Describe a method for unplugging a hole in a PC board that has become filled with solder.

Normal and Abnormal Voltages and Waveforms

2 . 1 GENERAL CONSIDERATIONS

It is helpful to consider dc voltage distributions in basic transistor circuits under conditions of normal and abnormal operation. An understanding of these facts assists the technician in analysis of trouble symptoms that occur in semiconductor networks and systems. Similarly, it is instructive to consider basic waveforms under normal operating conditions and the characteristic distortions that occur as a result of clipping, compression, frequency discrimination, and phase shift in semiconductor circuitry. Direct current voltages are usually measured with a dc voltmeter, although this measurement can also be made with a direct-coupled oscilloscope. Alternating current voltages are often measured with an ac voltmeter; however, experienced technicians generally prefer to measure peak-to-peak voltage values with an oscilloscope. This preference is based on the ability of an oscilloscope to display the waveshape that is associated with the peak-to-peak voltage value. For some readers, the topics discussed in this chapter will be a review and refresher; for others, these data will be new material and and a stepping stone to analysis of trouble situations in semiconductor circuitry.

2 . 2 VOLTAGE POLARITIES IN BASIC SEMICONDUCTOR CIRCUITS

Germanium and silicon bipolar (junction) transistors are manufactured in PNP and NPN types. A PNP transistor operates with its collector

negative with respect to its emitter, and an NPN transistor operates with its collector positive with respect to its emitter. Basic transistor circuits can be classified in two-battery arrangements and one-battery arrangements, as shown in Fig. 2–1. These are basic amplifier configurations. Each of the transistors in these examples is forward-biased. A PNP transistor is forward-biased if its base is negative with respect to its emitter. On the other hand, an NPN transistor is forward-biased if its base is positive with respect to its emitter. In these examples, the emitter of each transistor is grounded. Therefore, the collectors of the PNP transistors are negative with respect to ground, and the collectors of the NPN transistors are positive with respect to ground. The bases of the PNP transistors are negative with respect to ground, and the bases of the NPN transistors are positive with respect to ground.

Because the batteries are returned to the emitter terminal in Fig. 2–1(a) and (b), these configurations are called common-emitter (CE) circuits. Similarly, because the battery is returned to the emitter terminal in Figs. 2–1(c) and 2–1(d), these configurations are also called

Figure 2–1. Battery polarities in common-emitter configuration. **(a)** Two-battery PNP transistor circuit; **(b)** Two-battery NPN transistor circuit; **(c)** One-battery PNP transistor circuit; **(d)** One-battery NPN transistor circuit.

common-emitter circuits. In addition, they are examples of grounded-emitter circuits. The term *common ground* should not be confused with the term *common emitter*. To summarize briefly, a common ground is the point in a semiconductor circuit that is connected to the chassis of the equipment, or to an equivalent ground bus. On the other hand, a common-emitter circuit is one in which the voltage source(s) for collector and base are returned to the emitter terminal.

Next, with reference to Fig. 2–2, the batteries are returned to the base terminal of the transistor in the common-base (CB) configuration. In the examples of Figs. 2–2(c) and (d), the V_{cc} return is not made directly to the base terminal, but via resistors R2 and R3. We observe in each example of Fig. 2–2 that the input ac voltage is applied effectively between emitter and base, and that the output ac voltage is taken effectively between collector and base. These configurations are also examples of grounded-base circuits. The PNP transistors operate with base positive and collector negative with respect to base (and ground), whereas the NPN transistors operate with base negative and collector positive with respect to base (and ground).

Figure 2–2. Battery polarities in common-base configuration. **(a)** Two-battery PNP transistor circuit; **(b)** Two-battery NPN transistor circuit; **(c)** One-battery PNP transistor circuit; **(d)** One-battery NPN transistor circuit.

Figure 2–3. Battery polarities in common-collector configuration. **(a)** Two-battery PNP transistor circuit; **(b)** Two-battery NPN transistor circuit; **(c)** One-battery PNP transistor circuit; **(d)** One-battery NPN transistor circuit.

Another basic mode of operation employs the common-collector (CC) configuration, exemplified in Fig. 2–3. Battery return is made to the collector in the CC arrangement. These are also examples of grounded-collector circuits. We shall observe that the input ac voltage is applied effectively between base and collector, and that the output ac voltage is taken effectively between emitter and collector. The PNP transistors operate with base negative with respect to emitter and with collector negative with respect to emitter. On the other hand, the NPN transistors operate with base positive with respect to emitter and with collector positive with respect to emitter.

It is instructive here to consider basic examples of voltages with respect to ground, as depicted in Fig. 2–4. A common-emitter configuration does not necessarily employ a grounded-emitter connection. Thus, in (a), a PNP transistor operates in the CE mode with the negative terminal of V_{cc} grounded. In turn, all of the circuit voltages are positive with respect to ground. On the other hand, the collector voltage is negative with respect to the emitter terminal, and the base is negative with respect to the emitter terminal. Next, in (b),

an NPN transistor operates in the CE mode with the positive terminal
of V_{cc} grounded. In turn, all of the circuit voltages are negative with
respect to ground. On the other hand, the base and collector voltages
are positive with respect to the emitter terminal.

In Fig. 2–4(c) a PNP transistor operates in the CC mode with
the positive terminal of V_{cc} grounded. Accordingly, all of the circuit
voltages are negative with respect to ground. Moreover, the base and
collector voltages are negative with respect to the emitter terminal.
On the other hand, the base and collector voltages are positive with
respect to the collector. Next, in (d), an NPN transistor operates in
the CC mode with the negative terminal of V_{cc} grounded. Accordingly,
all of the circuit voltages are positive with respect to ground. More-
over, the base and collector voltages are positive with respect to the
emitter terminal. On the other hand, the base and collector voltages
are negative with respect to the collector.

Figure 2–4. Examples of voltages with respect to ground. **(a)** PNP transistor in
CE configuration, battery negative terminal grounded; all voltages are positive
with respect to ground; **(b)** NPN transistor in CE configuration, battery positive
terminal grounded; all voltages are negative with respect to ground; **(c)** PNP
transistor in CC configuration, battery positive terminal grounded; all voltages are
negative with respect to ground; **(d)** NPN transistor in CC configuration, battery
negative terminal grounded; all voltages are positive with respect to ground.

Figure 2–5. Examples of dc voltage distributions. **(a)** Circuit voltages measured with respect to ground; **(b)** Same circuit voltages as in (a), but measured with respect to emitter; **(c)** Emitter is more positive than base; transistor is forward-biased; **(d)** Emitter is more negative than base; transistor is forward-biased.

As depicted in Fig. 2–5, a germanium transistor operating in a class-A amplifier circuit employs forward bias of approximately 0.3 volt. Referring to (a), we see that the negative terminal of V_{cc} is grounded. Thus, all voltages in the circuit are positive with respect to ground. Since the emitter is more positive than the base, the transistor is forward-biased. The value of forward bias is equal to the difference between 2.4 and 2.1 volts, or 0.3 volt. Next, with reference to (b), the transistor terminal voltages are negative with respect to the emitter (are less positive than the emitter voltage). This principle of voltage distribution is again exemplified in (c); in this case, the emitter and base voltages are much higher than the collector voltage, but their difference is 0.3 volt, and the transistor is forward-biased. Finally, a forward-biased NPN transistor is depicted in (d) with its emitter terminal at a higher negative potential than its base terminal.

2.3 CHANGES IN DC DISTRIBUTION OWING TO TRANSISTOR DEFECTS

Defective transistors are usually pinpointed by changes in dc voltage distribution. As an illustration, Fig. 2–6(a) shows a normal distribution

Figure 2–6. Effect of an open collector junction. **(a)** Typical dc voltages for a silicon PNP transistor; **(b)** Collector voltage rises to V_{CC} when collector junction is open; **(c)** Typical dc voltages for a silicon NPN transistor; **(d)** Collector voltage rises to V_{CC} when collector junction is open.

of dc voltages in a PNP silicon transistor circuit, and Fig. 2–6(b) shows the change in distribution caused by an open collector junction. In this situation, the collector voltage rises to the V_{CC} value, and the base-emitter bias voltage remains virtually unchanged. Next, Fig. 2–6(c) shows a normal distribution of dc voltages in an NPN silicon transistor circuit, and Fig. 2–6(d) shows the change in distribution resulting from an open collector junction. As before, the collector voltage rises to the V_{CC} value, and the base-emitter bias voltage remains virtually unchanged. A silicon transistor has a typical forward bias of 0.6 volt for class-A operation. This is approximately twice the bias voltage for a germanium transistor in class-A operation.

Next, consider the changes in dc voltage distribution caused

by an open emitter junction. With reference to Fig. 2–7, a normal dc-voltage distribution for a germanium PNP transistor is depicted in (a), and the corresponding distribution change owing to an open emitter junction is shown in (b). In this situation, both the collector and base voltages rise to the V_{cc} potential. Similarly, a normal dc-voltage distribution for a germanium NPN transistor is indicated in (c), and the corresponding distribution change owing to an open emitter junction is seen in (d). As before, the collector and base voltages rise to the V_{cc} potential. When a junction is open, there is no current flow across the junction. This fact would be obvious in measurements of collector, base, or emitter currents in Figs. 2–7(b) or (d). On the other hand, professional technicians prefer to employ voltage

Figure 2–7. Effect of an open emitter junction. **(a)** Typical dc voltages for a germanium PNP transistor; **(b)** Collector and base voltages rise to V_{cc} when emitter junction is open; **(c)** Typical dc voltages for a germanium NPN transistor; **(d)** Collector and base voltages rise to V_{cc} when emitter junction is open.

measurements in the vast majority of trouble situations, because current measurements are comparatively difficult to make. This topic is discussed in greater detail subsequently.

A common transistor fault is called "leakage," or a "leaky collector junction." In other words, the junction is neither open-circuited nor short-circuited, but resistive; this leakage resistance has the effect of permitting excessive reverse-current flow. A normal dc-voltage distribution for a silicon NPN transistor is shown in Fig. 2–8(a), and the corresponding distribution change owing to a leaky collector junction is shown in (b). Junction leakage causes the base-emitter forward bias to increase, and causes the collector-emitter voltage to decrease. Since the forward bias voltage is excessive, the collector current is excessive. In turn, most of the V_{cc} voltage appears as an

(a) (b)

(c) (d)

Figure 2–8. Effect of a leaky collector junction. **(a)** Typical dc voltages for a silicon NPN transistor; **(b)** Collector voltage decreases and base voltage increases when collector junction is leaky; **(c)** Typical dc voltages for a silicon PNP transistor; **(d)** Collector voltage decreases and base voltage increases when collector junction is leaky.

Figure 2–9. Effect of a short-circuited emitter junction. **(a)** Typical dc operating voltages for a silicon NPN transistor; **(b)** Base voltage becomes zero and collector voltage equals V_{CC} when the emitter junction is short-circuited.

abnormal IR drop across R_L. Next, the corresponding normal and abnormal dc-voltage distributions for a silicon PNP transistor are depicted in Figs. 2–8(c) and (d).

Sometimes an emitter junction becomes short-circuited. In this situation, the base voltage assumes the same value as the emitter voltage (zero in the example of Fig. 2–9). In effect, the transistor has been changed into a reverse-biased diode, and the collector voltage rises to the V_{cc} potential. There is current flow through R1 through the short-circuited emitter junction to ground. However, this current cannot branch into the collector circuit because the emitter junction is lacking. As a practical troubleshooting note, a somewhat similar abnormal dc-voltage distribution can also be caused by an open connection to the base terminal, except that the base voltage that is measured will either be equal to V_{cc}, or will be a small positive voltage, depending upon which end of the open connection is being tested. It should also be pointed out that the base will be "floating" and the voltage value that is measured at the base will depend considerably upon the internal resistance of the voltmeter that is used. A low-resistance meter will indicate practically zero volts because of the very high reverse resistance of the collector junction.

2.4 RANGES OF TRANSISTOR OPERATING VOLTAGES

Typical operating voltage values have been discussed for germanium and silicon PNP and NPN transistor types. In practice, considerably

different values may be measured in various circuits, and the technician may suppose that trouble situations are indicated. Therefore, it is essential to recognize the normal ranges of transistor operating voltages. These ranges are shown in Fig. 2–10. Thus, V_{BE} for a small-signal PNP germanium transistor ranges from 0.05 to 0.3 volt, and V_{CE} ranges from 2 to 9 volts. Next, V_{BE} for a small-signal NPN silicon transistor ranges from 0.6 to 5 volts, and V_{CE} ranges from 10 to 60 volts. Again, V_{BE} for a medium-power NPN silicon transistor ranges from 2 to 8 volts, and V_{CE} ranges from 30 to 70 volts. Finally, V_{BE} for a high-power silicon transistor ranges from 2 to 6 volts, and V_{CE} ranges from 70 to 400 volts. Therefore, distinction between normal and abnormal operating voltages must usually be made by consulting the service data for the equipment under test.

Previous mention has been made of operating voltages for class-A operation of transistors. Many power output stages operate in class B. A class-B stage may be zero-biased (base and emitter voltages equal), or a very small forward bias may be employed. A transistor that is slightly forward-biased is said to operate in class AB. Transistors in oscillator circuits often operate in class C. As exemplified in Fig. 2–11, the base-emitter junction is reverse-biased in class-C operation. Although the transistor would *seem* to be cut off, this is not actually the case, inasmuch as base current and collector current flows in brief pulses. In other words, current flows on the peaks of the ac voltage in the circuit. A voltmeter can only measure the average value of the pulsating dc voltage at the base and emitter

V_{CE}
2 to 9V

V_{BE}
0.05 to 0.3V

(a) Small - signal PNP Germanium type

V_{CE}
10 to 60V

V_{BE}
0.6 to 5V

(b) Small - signal NPN Silicon type

V_{CE}
30 to 70V

V_{BE}
2 to 8V

(c) Medium - power NPN Silicon type

V_{CE}
70 to 400V

V_{BE}
2 to 6V

(d) High - power NPN Silicon type

Figure 2–10. Ranges of transistor operating voltages.

Figure 2-11. Oscillator transistor is reverse-biased in normal operation.

terminals; therefore, it can be falsely concluded that the transistor is cut off and that the circuit is inoperative. Therefore, dc voltage measurements may need to be supplemented by oscilloscope tests, as explained next.

2.5 BASIC WAVEFORMS AND DISTORTED WAVESHAPES

A sine wave is the fundamental type of waveform; as shown in Fig. 2–12(a), it has a positive-peak voltage, a negative-peak voltage, a root-mean-square (rms) voltage, and a peak-to-peak voltage. These positive- and negative-peak voltages are equal, and the peak-to-peak voltage is double the peak voltage. The rms voltage is indicated by conventional ac voltmeters, and it is equal to 0.707 of the peak voltage. Thus, the peak voltage is equal to 1.414 times the rms voltage, and the peak-to-peak voltage is equal to 2.83 times the rms voltage. In oscilloscope test procedures, the peak-to-peak voltage of a waveform is generally measured, and compared to the value specified in the equipment service data.

Next, it is instructive to consider some basic kinds of distortion that are found in malfunctioning equipment. If a transistor is driven into cutoff or into saturation, the output waveform will be clipped, although the input waveform is a true sine wave. As an illustration, Fig. 2–12(b) shows an example of positive peak clipping. This type of distortion occurs in the CE configuration if the base-emitter bias is subnormal. If the base-emitter bias were zero, the entire positive half-cycle would be clipped, and the output waveform

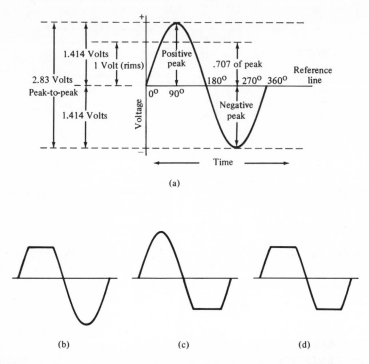

Figure 2–12. Basic sine wave and typical distortions. **(a)** Voltage values of the ideal sine waveform; **(b)** Positive-peak clipping; **(c)** Negative-peak clipping; **(d)** Positive- and negative-peak clipping.

would consist of a series of half-sine waves. Next, Fig. 2-12(c) shows an example of negative peak clipping. This type of distortion occurs in the CE configuration if the base-emitter bias is abnormal. Finally, suppose that the transistor is optimally biased for class-A operation, but that excessive drive signal is applied. In such a case, the transistor is driven both into saturation and past cutoff, with the results that both peaks of the output waveform are clipped, as depicted in Fig. 2–12(d).

Peak compression (Fig. 2–13) is a less abrupt form of distortion than peak clipping. It is commonly encountered in malfunctioning audio amplifiers. The severity of waveform distortion in audio systems is commonly expressed as a percentage of harmonic distortion, and is measured with a harmonic-distortion meter. High-fidelity reproduction is generally considered to entail less than 1 percent harmonic distortion. Another common type of audio waveform distortion is called *crossover distortion*, as seen in Fig. 2–14. The more severe examples of crossover distortion are clearly shown in an oscilloscope

display. Less severe crossover distortion is less apparent in a scope pattern, but is readily measurable with a harmonic-distortion meter. We shall find that the percentage of harmonic distortion that is present in an output waveform that has crossover distortion increases steadily as the amplitude of the output waveform is reduced. This relation is based on the fact that the distortion occurs in the vicinity of the zero axis.

Next, it is instructive to consider the chief types of square-wave distortion, depicted in Fig. 2–15. Square-wave distortion entails

<div align="center">(a)</div>

<div align="center">(b)</div>

<div align="center">(c)</div>

<div align="center">(d)</div>

<div align="center">(e)</div>

<div align="center">(f)</div>

Figure 2–13. Compression distortion of a sine wave. **(a)** One percent harmonic distortion; **(b)** Three percent harmonic distortion; **(c)** Five percent harmonic distortion; **(d)** Ten percent harmonic distortion; **(e)** Fifteen percent harmonic distortion; **(f)** Twenty percent harmonic distortion.

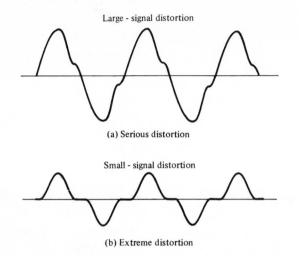

Figure 2–14. Crossover distortion of a sine wave.

both amplitude changes in various harmonic components, and phase shifts of various harmonic components. Many of the characteristics of square-wave response can be understood from the viewpoint of synthesis of a square wave from an array of sine waves, as exemplified in Fig. 2–16. As an illustration, if the fundamental frequency of a square wave is attenuated, it is apparent that the top of the waveform will become concave. On the other hand, if the fundamental frequency of a square wave is accentuated, it is apparent that the top of the waveform will become convex. Or, in case the higher harmonics are attenuated, it is evident that the corners of the waveform will become rounded.

Pulse waveform distortions are closely related to square-wave distortions. Basic pulse distortion terminology is illustrated in Fig. 2–17. It is instructive to recognize that distortion in the rapidly rising interval of a pulse (or square) waveform is associated with the high-frequency response of the circuit, whereas distortion in the flat-top interval is associated with the low-frequency response of the circuit. Overshoot is defined as the amplitude of the first maximum excursion in a pulse waveform beyond the 100 percent amplitude level, expressed as a percentage of this 100 percent amplitude. Preswing is simply "downward" overshoot. Ringing denotes a damped oscillation following the rise or fall of a pulse or square wave. It is caused by circuitry that has less than critical damping (Q value greater than 1). Tilt denotes the extent of upward or downward slope in the top of a pulse or square waveform.

Frequency distortion,
(amplitude reduction of
low frequency component)
no phase shift

Low frequency boost,
(accentuated fundamental)
no phase shift

High frequency loss
no phase shift

Low frequency phase
shift

Low frequency loss and
phase shift

High frequency loss and
low frequency phase shift

High frequency loss
and phase shift

Damped oscillation

Low frequency phase
shift

(a)

(b)

Figure 2–15. **(a)** Basic types of square-wave distortion; **(b)** Square-wave generator. *(Courtesy of RCA)*

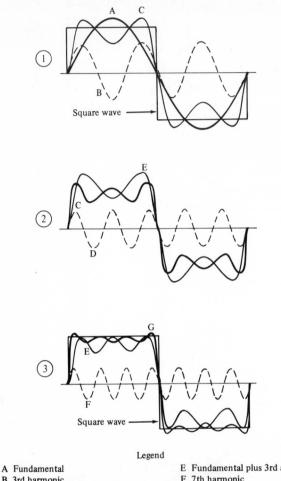

Legend

A Fundamental	E Fundamental plus 3rd and 5th harmonics
B 3rd harmonic	F 7th harmonic
C Fundamental plus 3rd harmonic	G Fundamental plus 3rd, 5th and 7th harmonics
D 5th harmonic	

Figure 2–16. Synthesis of a square wave from a fundamental sine wave and its harmonics.

In high-speed pulse circuitry, such as that encountered in digital systems, a characteristic type of pulse distortion occurs, as shown in Fig. 2–18. The top of the output waveform displays plateaus that correspond to the transition from unsaturated to saturated operation of a transistor. In other words, when a bipolar transistor is driven into heavy conduction, a point is reached at which the base voltage is greater than the collector voltage. At this point, both junctions are forward-biased, and the transistor is said to be saturated.

This condition is associated with accumulation of charge carriers in the base region, and a significant time interval is required to discharge these carriers after the drive voltage is removed. This interval is denoted by the storage time t_s in Fig. 2–18. Although not strictly a form of pulse distortion, a delay time t_d is also imposed between input and output signals, as seen in the diagram.

Another class of waveforms is called *frequency-response curves*, as exemplified in Fig. 2–19. A frequency-response curve displays the output voltage-versus-frequency characteristic of a tuned circuit or of a combination of tuned circuits. In the present example, the frequency response of an RF tuner in a television receiver is displayed. A channel-4 response curve, for example, encompasses a frequency progression from 66 to 72 MHz. A frequency-response curve shows the circuit transfer characteristic for a frequency-modulated (sweep-frequency) signal provided by a sweep-frequency generator. Key frequency points along a response curve are indicated by markers, also called "pips" or "birdies"; the picture-carrier and sound-carrier frequencies are marked on the curve in Fig. 2–19. Markers are produced by mixing the output from a conventional signal generator with the output from a sweep-frequency generator. In turn, visible beats appear on the response curve where the marker frequency zero-beats with the sweep frequency. Common distortions of frequency-response curves include inadequate bandwidth, excessive bandwidth, unequal response at low- and high-frequency ends, exces-sive peaking, and excessive sagging.

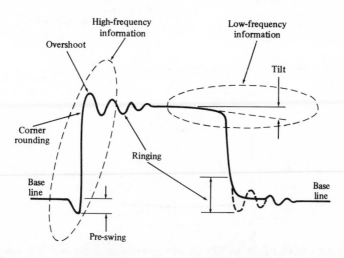

Figure 2–17. Pulse distortion terminology.

t_d = pulse delay time
t_r = pulse rise time
t_p = maximum amplitude duration time
t_s = storage time
t_f = pulse fall time

Figure 2–18. Pulse distortion in high-speed transistor switching cycle.

Figure 2–19. Example of an RF frequency-response curve with picture- and sound-carrier markers.

Still another class of waveforms is termed *Lissajous figures* (patterns), as exemplified in Fig. 2–20. In this type of display, related signals, such as audio input and output signals, are applied to the vertical and horizontal amplifiers of an oscilloscope. As an illustration, an audio amplifier is tested for distortion by driving it with the sine-

(a)

A. No overload distortion, no phase shift.
B. Overload distortion, no phase shift.
C. Driving into saturation and past cut-off, no phase shift.
D. Phase shift.
E. Overload distortion and phase shift
F. Phase shift, driving into cut-off and into saturation.
G. Crossover distortion.

(b)

Figure 2–20. A basic method of distortion analysis. **(a)** Test setup; **(b)** Pattern evaluations.

wave output from an audio oscillator. In turn, the input signal to the amplifier is applied to the vertical-input terminals of the scope, and the output signal from the amplifier is applied to the horizontal-input terminals of the scope. The resulting patterns display various forms of distortion (if present), such as overload, phase shift, cutoff, saturation, amplitude nonlinearity, and crossover distortion. Remember that slight amounts of distortion are virtually imperceptible in oscilloscope displays, although serious distortion is clearly visible.

REVIEW QUESTIONS

1. Explain the meaning of "dc voltage distribution."
2. Why do technicians generally prefer to measure peak-to-peak voltage values with an oscilloscope?
3. Define a common-emitter circuit; a common-base circuit; a common-collector circuit.
4. What is the significance of a common-ground point in a circuit?
5. May a common-emitter circuit have an arbitrary common-ground point?
6. State a typical forward-bias value for a germanium transistor; for a silicon transistor.
7. Do transistor defects usually result in dc-voltage distribution changes?
8. Discuss the effect on circuit voltages of an open collector junction; of an open emitter junction.
9. What are the symptoms of a leaky collector junction?
10. How would it be determined that an emitter junction is short-circuited?
11. Explain what is meant by a "floating" electrode (or terminal).
12. State the normal ranges of transistor operating voltages for a small signal germanium transistor.
13. What are the normal ranges of transistor operating voltages for a small-signal silicon transistor?
14. Define class-A operation of a transistor; class-B operation; class-C operation.
15. Describe the most fundamental type of waveform.

16. Explain the relations among rms, peak, and peak-to-peak voltages for a sine wave.
17. What is meant by a "cutoff" condition in a transistor; a "saturated" condition?
18. How is waveform distortion measured?
19. State the chief types of square-wave distortion.
20. Discuss delay time between the input and output signals of a transistor.
21. Describe the meaning of storage time in a transistor.
22. How is a frequency marker produced on a frequency-response curve?
23. Explain how Lissajous patterns are developed.
24. State the forms of distortion that can be displayed by Lissajous patterns.

Troubleshooting Basic
Solid-state Amplifiers

3.1 SURVEY OF BASIC AMPLIFIERS

Solid-state troubleshooting procedures often involve localization of device or component defects in various types of amplifiers. Therefore, a technician needs a good understanding of amplifier malfunctions and circuit-testing techniques. Figure 3–1 shows the appearance of a typical solid-state amplifier. Most amplifiers have more than one stage of amplification, and the general plan of a multistage amplifier is represented by a block diagram, as exemplified in Fig. 3–2. A block diagram is sectionalized; in other words, it shows the principal sec-

Figure 3–1. Appearance of a solid-state high-fidelity audio amplifier. *(Courtesy of Radio Shack, a Tandy Corporation Company)*

Figure 3–2. Typical block diagram for an audio amplifier.

tions of the amplifier, without any circuit details. Next, a partial schematic diagram has the typical data shown in Fig. 3–3. Observe that the normal dc operating voltages are specified at each semiconductor terminal. This is an essential troubleshooting aid, as explained in greater detail subsequently.

The amplifier depicted in Fig. 3–3 employs three stages; the output stage comprises two transistors operating in push-pull. Bipolar (junction) transistors are utilized. Transistors Q1 and Q2 operate in the common-emitter mode, whereas Q3 and Q4 operate in the common-collector (emitter-follower) mode. The output stage provides complementary push-pull output to the speaker. In other words, Q3 amplifies the negative half-cycle of the output waveform, whereas Q4 amplifies the positive half-cycle of the output waveform. Note that the output audio signal is taken from the emitters of Q3 and Q4. Also, the audio-output signal is fed back via R4 to Q2, and C5, C6, and R13 to Q1. These are negative-feedback loops.

It is helpful to observe the basic characteristics of bipolar transistor amplifier stages in the three fundamental modes, as shown in Fig. 3–4. These are typical characteristics, and they will vary somewhat, depending upon the variety of transistor utilized and upon the value of the collector load resistor. If the internal resistance of the generator (driving source) is greater or less than 1 kilohm, this will also affect the characteristic values to some extent. Observe that the common-emitter (CE) configuration inverts the waveform polarity from input to output (shifts the waveform phase 180 degrees). On the other hand, the common-base (CB) and common-collector (CC) configurations do not invert the waveform polarity.

Of the three configurations in Fig. 3–4, the CE stage has the most nearly equal input and output resistances. It also has the highest power gain. Therefore, the CE configuration is generally used in cascaded arrangements. In other words, maximum power transfer is obtained when input and output resistances are the same (matched). Observe next that the CC or emitter-follower configuration has the lowest output resistance. Therefore, maximum power transfer is ob-

Figure 3–3. A partial schematic diagram for an audio amplifier.

tained by using a CC stage to drive the speaker, inasmuch as a speaker has comparatively low input resistance. Note that the power gain is given by the product of the voltage gain and the current gain. For example, a CC stage has a voltage gain of 1, a current gain of 36, a power gain of 36, and a corresponding dB gain of approximately 15. Gain measurements are often made by technicians in amplifier trouble-shooting procedures, as explained in greater detail subsequently.

Next, observe the FET preamplifier configuration depicted in Fig. 3–5. Junction field-effect transistors (JFET's) are utilized in the amplifier stages and also in the base and treble tone-control circuits. It is helpful to observe the basic characteristics of FET amplifier stages in the three fundamental modes, as shown in Fig. 3–6. The common-source configuration is most widely used. With reference to Fig. 3–5, FET preamplifier stages Q1 and Q2 provide impedance matching and amplification for a magnetic phono cartridge. C8 provides high-frequency rolloff; in other words, it provides equal amplification by the stage at all frequencies in the audio passband. Resistor R2 is a source-bias resistor, and is bypassed at audio frequencies by C1 to prevent signal degeneration. Note that Q1 has its gate biased nega-tively with respect to its source—it is an N-channel FET; if a P-

channel FET were employed, a positive gate-bias voltage would be used.

Observe in Figure 3–5 that R1 and R12, in conjunction with the FET input impedance, provide the required impedance match for a magnetic cartridge. C5 is a coupling capacitor, and R3 is the dc return for Q2 gate current. R6 and C3 provide low-frequency boost by partially shunting the high audio frequencies to ground. For Q2, R4/C2 is the source-bias/ac bypass arrangement. R8 is the drain load, and C7/R11 is the af decoupler. R7 provides current limiting, R5 is a terminating load, and C4 couples the output from Q2 to the next stage. Q3 and R13 provide a proper impedance match for a crystal pickup device, which can be connected to phono jacks J2, J3, and/or J4. R13 is the dc return for the Q3 gate current. R14/C9 is a source-bias/ac bypass, R17/C10 the af decoupler, and R16 the drain load. Alternating current stability is obtained in this stage by inclusion of unbypassed R15, which provides degenerative feedback to the gate input. C11 provides ac coupling from Q3 to the Q4 tone-control stage.

AC Input — Amplifier output voltage

Voltage gain: 270 times
Current gain: 35 times
Power gain: 40dB
Input resistance: 1.3K
Output resistance: 40K
(for generator internal resistance of 1K)

(a) Common emitter.

AC Input — Amplifier output voltage

Voltage gain: 380 times
Current gain: 0.98
Power gain: 26dB
Input resistance: 34 ohms
Output resistance: 1 meg
(for generator internal resistance of 1K)

(b) Common base.

AC Input — Output voltage

Voltage gain: 1
Current gain: 36 times
Power gain: 15dB
Input resistance: 350K
Output resistance: 500 ohms
(for generator internal resistance of 1K)

(c) Common collector.

Figure 3–4. Basic amplifier configurations and typical operating characteristics.

Figure 3–5. An FET preamp with FET bass and treble tone-control circuits. (Adapted from a Motorola or

Transconductance: 5000μmhos
Power gain: 17 dB (50 times)
Input resistance: Very high
Output resistance: 20K
(For generator internal resistance
of 500 ohms)

(a) Common source

Voltage gain: 1.8
Input resistance: 240 ohms
Output resistance: High
(For generator internal resistance
of 500 ohms)

(b) Common gate

Voltage gain: 0.5
Input resistance: 2M
Output resistance: 240 ohms
(For generator input resistance
of 500 ohms)

(c) Common drain

Figure 3–6. Basic FET amplifier configurations and typical operating conditions. *(Courtesy of RCA)*

As seen in Fig. 3–5, the RC network in the gate of Q4 provides both bass and treble tone control for the preamplifier. C14/ R35/C15 form a low-frequency boost network that attenuates the higher audio frequencies to an extent that is dependent upon the setting of R35. R18/R19/R34/C12/C13 form a low-frequency attenuator, or high-frequency boost network that attenuates the low frequencies to an extent that is dependent upon the setting of R34. R23 is the source-bias resistor for Q4. It is unbypassed and provides ac stability by degenerative feedback. R22 is the drain load, and C16/

R21 is the af decoupler. C17 ac-couples the output from Q4 to the following stage. Emitter follower Q5 provides a high load impedance for the volume control R36. This high input impedance characteristic is required to prevent variation in frequency response when the value of R36 is varied. R25 and R26 form a voltage divider that establishes correct operating bias for this stage. R24/C18 provides af decoupling. R27 is the emitter load from which the audio signal is ac-coupled by C20 to the base of Q6.

Transistor Q6 in Fig. 3–5 operates in the CE mode with a voltage gain of approximately 10. R28 is the base dc return, and in conjunction with R32 forms a voltage divider that establishes the required degenerative feedback for stable operation. C21/R29 is an af decoupler. R31 provides emitter bias and also contributes to ac and dc stability. Output is taken from collector load R30 and is direct-coupled to Q7. Transistor Q7 is another emitter follower that is used to provide a low-impedance output. Final audio output is taken from emitter load R33 and is ac-coupled through C22 to the output jack J5. It is evident that the ability to evaluate a schematic diagram and to recognize the circuit actions of the various networks is of great assistance in troubleshooting procedures.

3 . 2 METHODS OF IN-CIRCUIT TESTING

Direct current voltage measurements are basic in semiconductor troubleshooting procedures. It is good practice to use a voltmeter that has a high value of input resistance in order to avoid the possibility of erroneous measurements in high-resistance circuits. For example, a field-effect multimeter (TVM) such as that illustrated in Fig. 3–7 provides an input resistance of 15 megohms on all dc-voltage ranges. In return, the possibility of objectionable circuit loading is practically eliminated. Normal dc operating voltages are specified in receiver service data, as exemplified in Fig. 3–3. It is advisable to measure the supply voltage at the outset, inasmuch as an incorrect value of supply voltage will upset the dc-voltage distribution throughout the entire configuration. A bench power supply may be employed to provide a precise value of supply voltage during troubleshooting procedures. Figure 3–8 shows a typical bench power supply for solid-state servicing requirements.

In the example of Fig. 3–3, consider a trouble symptom consisting of distorted sound output. This symptom could be caused by a defective transistor, a faulty capacitor, an off-value thermistor, or an off-value resistor. Any of these defects will disturb the dc-voltage

Figure 3–7. A typical TVM with an input resistance of 15 megohms. *(Courtesy of Sencore, Inc.)*

distribution, with one exception. In other words, an open capacitor will not change dc-voltage values, although it can cause distorted sound output. Therefore, dc-voltage measurements are very likely to lead the technician to the trouble area, if not to the defective component. In case the faulty component is not pinpointed by dc-voltage measurements, supplementary tests can be made as required. In this example, trouble is tracked down to the driver stage; the emitter voltage at Q2 measures —0.1 volt, and the base voltage measures —0.28 volt. In turn, transistor Q2 falls under suspicion, as well as the resistors and capacitors in its base and emitter circuits. To minimize testing time in this situation, it is good practice to consider the "most likely defects."

Statistically, electrolytic capacitors are more likely to fail than

Figure 3–8. A typical bench power supply. *(Courtesy of Sencore Inc.)*

are transistors or resistors. Therefore, with reference to Fig. 3–3, initial suspicion falls upon C1 and C2. Electrolytic capacitors affect the dc-voltage distribution by becoming leaky or short-circuited. For example, if C1 becomes leaky, the base voltage at Q2 will increase. However, the foregoing measurements have shown that the base voltage at Q2 has decreased (from −0.85 volt to −0.28 volt). Accordingly, C1 can be eliminated from suspicion. Next, if C2 becomes seriously leaky, it will bring the emitter voltage at Q2 nearer to ground potential. Note that the foregoing measurements have shown that the emitter voltage has decreased (from −0.69 volt to −0.1 volt). In turn, it is logical to assume at this point that C2 is very leaky. This assumption is found to be correct, and when C2 is disconnected from the circuit board, its resistance measures approximately 1 ohm. Replacement of C2 restores amplifier operation to normal.

Next, consider an important type of in-circuit test called the *transistor turn-off test*. This is a simple quick check for determining whether a transistor has normal control action. Again with reference to Fig. 3–3, the technician may suspect that transistor Q2 is defective, and proceed to make a turn-off test. This is done as shown in Fig. 3–9. A temporary short circuit is applied between the base and emitter of

Figure 3–9. Example of a transistor turn-off test.

the transistor, and the resulting change in collector voltage is measured with a TVM. When the base and emitter of a bipolar transistor are brought to the same potential, the transistor normally cuts off. In turn, the collector voltage normally rises to the supply-voltage value (approximately —8 volts in this example). If the transistor happens to be defective, its collector voltage will not rise when the base and emitter terminals are short-circuited.

An associated type of in-circuit test is called the *transistor turn-on test*. This is another simple quick check for determining whether a transistor has normal control action. As before, with reference to Fig. 3–3, the technician may suspect that transistor Q2 is defective, and proceed to make a turn-on test. This is done as shown in Fig. 3–10. A 50-kilohm resistor is temporarily applied between the collector and base of the transistor, and the resulting change in col-

Figure 3–10. Turn-on test of transistor control action.

lector current is indicated by a TVM connected across the emitter resistor. When bias voltage is bled into the base of a bipolar transistor from its collector terminal, the transistor normally increases collector-current conduction. In turn, the voltage drop across the emitter resistor increases. If the transistor happens to be defective, its emitter voltage will not rise when the forward base bias is increased.

Next, it is instructive to consider the in-circuit measurement of resistance values. With reference to Fig. 3–11, three resistors are connected to the terminals of a bipolar transistor. Each of these resistors is shunted effectively by transistor junction resistance. Each junction has two resistance values: a low forward resistance and a very high reverse resistance. In turn, the apparent value of the 10-kilohm resistor, when measured in circuit, reads 26 ohms with emitter positive and base negative. On the other hand, the ohmmeter will read 10,000 ohms with emitter negative and base positive. Next, consider the apparent value of the 1.2-meg resistor when measured in-circuit. With collector positive and base negative, the ohmmeter will read 12,000 ohms. On the other hand, with collector negative and base positive, the ohmmeter will read 1.2 megohms. These polarity-dependent resistance readings are the result of transistor "turn-on" when a junction is forward-biased by the ohmmeter test voltage.

It follows from the previous examples that the values of the resistors depicted in Fig. 3–11 can be correctly measured in circuit, regardless of the test-voltage polarity, provided that the test voltage is sufficiently small that the transistor junctions cannot be "turned on." To meet this requirement, the ohmmeter must apply less than 0.1 volt across the circuit under test. Such ohmmeters are available for semiconductor troubleshooting. For example, the instrument illustrated in Fig. 3–7 has both a high-power and a low-power ohmmeter arrangement. In turn, the ohmmeter can be operated on its low-power function when in-circuit resistance measurements are to be made. In other words, the resistance value of each resistor in Fig. 3–11 can be accurately measured in circuit with a low-power ohmmeter, regardless of the test-voltage polarity that is applied. Therefore, a low-power ohmmeter is of considerable assistance when one is troubleshooting semiconductor circuitry.

Consider next the possibility of a defective transistor in the circuit of Fig. 3–11. For example, the transistor may be "shorted" (have defective junctions with very low forward and reverse resistance values). In such a case, a low-power ohmmeter indicates very low resistance values for the 10-kilohm and the 1.2-megohm resistors measured in circuit. The experienced troubleshooter evaluates these subnormal resistance readings as evidence of a "shorted" transistor, because it

Figure 3–11. Example of polarity-dependent in-circuit resistance readings.

is very seldom that two resistors in a circuit both decrease greatly in value. Again, consider the possibility of an "open" transistor. This defect will also be indicated by an ohmmeter test—all three resistors in Fig. 3–11 read correct values on a low-power ohmmeter, *and will also read correct values on a high-power ohmmeter.* In turn, the logical conclusion is that the transistor is "open."

Of course, it is not always possible to measure the value of a resistor in circuit. As an illustration, with reference to Fig. 3–3, R7 is shunted by R8. Therefore, one end of R7 must be disconnected before its resistance value can be measured. Note that after R7 has been disconnected from the circuit, the value of R8 can be measured in circuit with a low-power ohmmeter. To temporarily disconnect a component from a PC conductor, the technician may cut across the conductor with a razor blade. The conductor is thereby open-circuited. After the desired test has been completed, the connection is restored by melting a small drop of solder across the razor slit (cut with a razor blade).

In general, the troubleshooting advantages of a low-power ohmmeter apply only to junction (bipolar) transistor circuitry. Field-effect transistors do not have junctions between drain and source, and this fact may defeat the purpose of a low-power ohmmeter. Before attempting to make in-circuit resistance measurements in drain-source circuitry, the technician must determine whether the FET is a depletion type or an enhancement type of device. The distinction is depicted in Fig. 3–12. Note that a depletion-type FET has a low value of drain-source resistance in the vicinity of zero bias. On the other hand, an enhancement-type FET has an extremely high value of drain-source resistance in the vicinity of zero bias. In other words, in-circuit resistance measurements in FET drain-source circuitry are feasible, provided that an enhancement-type FET is employed. Otherwise, in-circuit drain-source resistance measurements are impractical.

Figure 3–12. Drain-source conduction characteristics for depletion-type and enhancement-type FET's.

Gate-circuitry resistance measurements are always possible with a low-power ohmmeter, whether a JFET or a MOSFET is employed. In other words, a JFET has a conventional junction that will not be "turned on" by the low test voltage of a low-power ohmmeter. Again, a MOSFET has an insulated gate construction that is practically an open circuit, regardless of ohmmeter polarity. When testing in N-channel JFET circuitry, the resistors in the gate network can be checked in-circuit with a low-power ohmmeter. On the other hand, the resistors in the drain-source network cannot be checked in circuit.

Therefore, the technician may "razor" the drain PC conductor, or both the drain and source PC conductors for purposes of test. With the drain and source PC conductors "razored," an out-of-circuit test can be made of the FET, without unsoldering the device from the circuit board.

Another important type of in-circuit test consists of bridging a suspected open capacitor with a known good capacitor, to determine whether normal operation is resumed. As an illustration, if the sound output is weak from the configuration in Fig. 3–5, it may be suspected that C1 or C2 is open, or has lost a major portion of its capacitance. In such a case, the technician often shunts the suspected capacitor temporarily with a new capacitor of suitable value and rating, to determine whether normal operation is resumed. If the test indicates that the suspected capacitor is indeed faulty, it is then removed from the circuit board and replaced. This bridging test is also made for suspected open- or low-capacitance coupling capacitors, such as C20 or C22 in Fig. 3–5. It is good practice to turn the amplifier off before a bridging capacitor is applied across terminals of a suspected open capacitor. In other words, *a large capacitor demands a sudden surge of current when it is inserted into a "live" circuit, and this surge occasionally causes transistor junction ratings to be exceeded, with resulting damage to the device.*

3 . 3 COLOR CODES

Knowledge of color codes is essential in practical semiconductor troubleshooting procedures. Unless a technician can read the color coding of resistors and capacitors, an excessive amount of time will be wasted in making unnecessary measurements. The resistor color code is shown in Fig. 3–13, and the capacitor color codes are shown in Fig. 3–14. The colored bands around the body of a color-coded resistor represent its value in ohms, with a certain tolerance. These colored bands are grouped toward one end of the resistor body. Starting with this end of the resistor, the first band represents the first digit of the resistance value; the second band represents the second digit; the third band represents the number by which the first two digits are multiplied. A fourth band of gold or silver represents a tolerance of ± 5 or ± 10 percent, respectively. The absence of a fourth band indicates a tolerance of ± 20 percent. The physical size of a composition resistor is related to its wattage rating. Size increases progressively as

CODE

COLOR	1st DIGIT	2nd DIGIT	MULTIPLIER
BLACK	0	0	1
BROWN	1	1	10
RED	2	2	100
ORANGE	3	3	1,000
YELLOW	4	4	10,000
GREEN	5	5	100,000
BLUE	6	6	1,000,000
VIOLET	7	7	10,000,000
GRAY	8	8	100,000,000
WHITE	9	9	1,000,000,000
GOLD	-	-	.1
SILVER	-	-	.01

TOLERANCE GOLD ± 5%
SILVER ± 10%
NO BAND ± 20%

EXAMPLES

YELLOW–4
VIOLET–7 } 47 x 1,000
ORANGE–1,000 } (or 47KΩ)

SILVER = TOLERANCE ± 10%

RED–2
RED–2 } 22 x 100,000
GREEN–100,000 } (or 2.2 MEG.)

NO BAND = TOLERANCE ± 20%

Figure 3–13. Color code for composition (carbon) resistors.

the wattage rating is increased. Thus, the diameters of $\frac{1}{2}$-watt, 1-watt, and 2-watt resistors are approximately $\frac{1}{8}$, $\frac{1}{4}$, and $\frac{5}{16}$ in., respectively.

Generally, only mica and tubular ceramic capacitors, used in modern solid-state equipment, are color-coded. The color codes differ somewhat among various capacitor manufacturers, but the codes shown in Fig. 3–14 are in very extensive use. The temperature coefficient is seldom important in a replacement capacitor for an audio amplifier. However, the temperature coefficient may be of great im-

MICA

CODE

COLOR	1st DIGIT	2nd DIGIT	MULTIPLIER	TOLER. %
BLACK	0	0	1	± 20
BROWN	1	1	10	–
RED	2	2	100	± 2
ORANGE	3	3	1,000	± 3
YELLOW	4	4	10,000	–
GREEN	5	5	–	± 5
BLUE	6	6	–	–
VIOLET	7	7	–	–
GRAY	8	8	–	–
WHITE	9	9	–	–
GOLD	–	–	.1	–
SILVER	–	–	.01	± 10

EXAMPLE

2 7 x 100
RED VIOL RED

MICA

GRN
± 5%

CHARACTERISTIC

2,700 pF ± 5%
OR .0027 μF

OBSERVE DIRECTION OF ARROW

WHT. OR BLK. DOT INDICATES MICA

CHARACTERISTIC – SEE NOTE 1 BELOW

(VALUE IN pF–SEE NOTE 3 BELOW)

TUBULAR CERAMIC

PLACE THE GROUP OF RINGS OR DOTS TO THE LEFT AND READ FROM LEFT TO RIGHT'

CODE

COLOR	1st DIGIT	2nd DIGIT	MULTIPLIER	TOLER. % 10 OR LESS	TOLER. % OVER 10
BLACK	0	0	1	± 20	± 20
BROWN	1	1	10	± 0.1	± 1
RED	2	2	100	–	± 2
ORANGE	3	3	1,000	–	± 2.5
YELLOW	4	4	10,000	–	–
GREEN	5	5	–	± 0.5	± 5
BLUE	6	6	–	–	–
VIOLET	7	7	–	–	–
GRAY	8	8	–	± 0.25	–
WHITE	9	9	–	± 1.0	± 10

EXAMPLE

TEMPERATURE COEFFICIENT– SEE NOTE 2 BELOW

TEMPERATURE COEFFICIENT– SEE NOTE 2 BELOW

ORG ORG BLK GRN
3 3 x 1 ± 5%

33pF ± 5%

(VALUE IN pF–SEE NOTE 3 BELOW)

NOTES: 1. The characteristic of a mica capacitor is the temperature coefficient, drift capacitance and insulation resistance. This information is not usually needed to identify a capacitor but, if desired, it can be obtained by referring to EIA Standard, RS-153 (a Standard of Electronic Industries Association.)

2. The temperature coefficient of a capacitor is the predictable change in capacitance with temperature change.and is expressed in parts per million per degree centigrade. Refer to EIA Standard, RS-198 (a Standard of Electronic Industries Association.)

3. The farad is the basic unit of capacitance, however capacitor values are generally expressed in terms of μF (micro-farad, .000001 farad) and pF (micro-micro-farad, .000001 μF); therefore, 1,000 pF = .001 μF, 1,000,000 pF = 1 μF.

Figure 3–14. The EIA standard capacitor color codes.

portance when one is replacing capacitors in high-frequency tuned circuits, as explained in greater detail subsequently.

3.4 PRINCIPLES OF SIGNAL TRACING

Signal tracing is another of the basic in-circuit test procedures used in semiconductor troubleshooting. An oscilloscope is generally used as the signal indicator, and the amplifier under test is usually driven by an audio oscillator. A typical oscilloscope is illustrated in Fig. 3–15, and an audio oscillator of good quality is pictured in Fig. 3–16. Signal-tracing tests are used to determine where the signal stops in case of a "dead amplifier" trouble symptom. For example, with reference to Fig. 3–5, the output from an audio oscillator may be applied at J1. A comparatively low-level signal voltage is used, such as 5 or 10 millivolts. If excessive input voltage is applied, the amplifier will overload and appear to distort the signal, even though it is functioning normally. To avoid excessive noise output from the amplifier during test procedures, it is common practice to disconnect the speaker(s) and to employ a resistor load instead. A power resistor of suitable

Figure 3–15. A basic high-performance oscilloscope. *(Courtesy of Hewlett Packard Company)*

Figure 3–16. A good-quality audio oscillator. *(Courtesy of Hewlett-Packard)*

resistance value and wattage rating is required to substitute properly for a speaker.

The oscilloscope should be used with a high impedance, low-capacitance probe, to avoid the possibility of loading high-impedance circuits, such as the gate circuits of Q1, Q2, or Q3 in Fig. 3–5. A test frequency of 1 kHz is suitable, and the oscilloscope controls are adjusted to display a pattern such as that illustrated in Fig. 3–17. The input signal is traced, starting at the gate of Q1, and then proceeding in turn to the drain of Q1, to the gate of Q2, to the drain of Q2, and so on. In normal operation, a greater vertical deflection is obtained at the drain than at the gate of a transistor. As signal tracing proceeds from one stage to the next, the vertical-gain control of the oscilloscope must be turned down to avoid off-screen deflection. Note that when an emitter follower, such as Q5, is tested, there is normally no gain (or a slight voltage loss) from base to emitter, as indicated in Fig. 3–4.

As an example of trouble location, consider the test results obtained if there is an open-circuited gate junction in Q4 (Fig. 3–5). In such a case, a sine-wave pattern of increasing amplitude is observed up to the drain of Q4. However, no pattern is found at the drain ter-

Figure 3–17. A typical sine-wave pattern observed during signal-tracing procedures.

minal of Q4, nor at any subsequent transistor. In turn, the technician concludes that the trouble involves either a defective FET, or some associated circuit defect, such as a short circuit in C16. To establish that the FET is defective, a dc-voltage check is made at the drain terminal; the presence of dc voltage is strong evidence that the FET is faulty. In turn, most technicians proceed to replace the FET without making further tests. However, an in-circuit transistor tester can be utilized, if desired, to confirm that the FET is defective before it is un-soldered from its circuit.

Another trouble symptom that can be easily analyzed by signal-tracing tests is "weak output." In this situation, the signal is traced through the amplifier as explained above, with particular attention to the progressive increase (or lack of increase) in pattern height from gate to drain of each transistor. For example, suppose that C19 has lost a substantial portion of its rated capacitance. In turn, a serious decrease in signal amplitude is found when the test probe is moved from the drain of Q4 to the base of Q5, even though the volume control is advanced to maximum. A supplementary test at the input terminal of the volume control shows practically no signal loss. Therefore, the conclusion would be made that C19 is defective, and the capacitor would be replaced without further tests.

3.5 BASIC DISTORTION ANALYSIS AND TESTS

Still another trouble symptom that is often analyzed by means of signal-tracing tests is "distorted output." It should be noted that all amplifiers distort to some extent, although the percentage of distortion may be so low that the output is considered to be distortionless. Although no industry standards have been established, it is generally agreed that high-fidelity performance entails a distortion value of less

than 1 percent. It is instructive to note that it is practically impossible to detect 1 percent distortion in an oscilloscope pattern. For example, the waveform illustrated in Fig. 3–17 has a 1 percent distortion, as measured with a harmonic-distortion meter. Visually, the pattern appears to be a perfect sine wave. If an amplifier develops 10 percent distortion, the associated sine waveform is obviously abnormal, as seen in Fig. 3–18(c). Even 5 percent distortion is clearly evident to the trained oscilloscope operator. A highly skilled technician can detect 3 percent distortion in a sine waveform. As a practical note, distortion analysis requires the use of a laboratory-type oscilloscope that has negligible inherent distortion.

The most usual cause of distortion is incorrect bias on a transistor. As an illustration, distortion in the output section of the amplifier depicted in Fig. 3–5 could be caused by leakage in C20. Leakage in C20 will result in bias shift both at the emitter of Q25 and at the base of Q6. Incorrect bias impairs the dynamic range of a transistor, and causes peak compression or clipping. This type of distortion is most prominent when the amplifier is driven to its full rated power output. After the source of distortion has been localized to a particular section of an amplifier by signal-tracing tests, the de-

(a) (b)

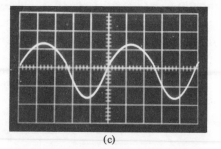

(c)

Figure 3–18. Examples of various percentages of distortion in sine-wave patterns. **(a)** Three percent harmonic distortion; **(b)** Five percent harmonic distortion; **(c)** Ten percent harmonic distortion.

fective component is then pinpointed by means of dc-voltage and re-sistance measurements, as explained previously. In other words, dis-tortion in the output section of the amplifier shown in Fig. 3–5 could be caused by collector-junction leakage in Q5, Q6, or Q7. A less likely cause is a substantial drift in value of a resistor in this section of the amplifier.

Signal-tracing procedures are useful for localizing the source of substantial distortion in the foregoing example, because the am-plifier does not employ feedback loops. On the other hand, the con-figuration shown in Fig. 3–3 includes two feedback loops, with the result that waveform analysis becomes involved. It is instructive to consider the basic factors that must be contended with in this situ-ation. Figure 3–19 shows a simple example of a feedback loop. Nega-tive feedback is utilized, with a part of the output signal fed back to the input of the amplifier. In turn, the amplitude of the input signal is effectively reduced, the gain of the amplifier is reduced, and the percentage of distortion of the amplifier is reduced. Note that negative feedback does not reduce distortion that might be present in the signal source. Negative feedback can only reduce distortion that is developed in the amplifier at any point that is included within the feedback loop. This reduction of distortion is accomplished by automatically predis-torting the input signal in a manner that tends to cancel out the distortion developed in the amplifier.

Next, observe the amplifier arrangement with two negative-feedback loops depicted in Fig. 3–20. This block diagram corresponds to the schematic diagram shown in Fig. 3–3. Feedback loop 1 returns part of the output signal into the driver stage with opposing phase to the input signal. Feedback loop 2 returns part of the output signal into the preamp stage with opposing phase to the input signal. In turn, if a defective component in the output stage causes substantial dis-tortion, a corresponding predistorted signal is fed back into the pre-amp stage and the driver stage. This feedback results in reduced

Figure 3–19. A simple example of a negative-feedback loop.

Figure 3–20. An amplifier arrangement with two negative-feedback loops.

distortion in the output signal; it also confuses signal-tracing procedures for localizing the source of distortion. Therefore, when one is troubleshooting a "distorted output" symptom in an amplifier arrangement such as that depicted in Fig. 3–20, it is generally advisable to forego waveform checks and to proceed at the outset with systematic dc-voltage and resistance measurements.

A preamplifier arrangement employing integrated circuits is shown in Fig. 3–21. Note that this configuration has two negative-feedback loops; one loop feeds back a portion of the output signal from

Figure 3–21. Typical integrated-circuit preamp configuration.

terminal 6 to terminal 7, and the other loop feeds back a portion of the output signal from terminal 1 to terminal 3. Troubleshooting procedures for this type of amplifier are basically the same as those for amplifiers that have conventional transistors. In other words, each integrated circuit functions as an amplifier, and the IC can be regarded as an elaborate form of transistor. Preliminary trouble localization can usually be made on the basis of signal-tracing tests. Pinpointing of defective components is generally accomplished by dc-voltage and/or resistance measurements. If an IC is apparently defective, it cannot be conclusively checked with ordinary service-type equipment. Accordingly, a substitution test is required.

3.6 SERVICING PRECAUTIONS

When an amplifier has a "weak-output" trouble symptom, the technician may observe that the output level can be increased by applying abnormal voltage to the amplifier. In other words, if the rated supply voltage for the amplifier is 12 volts, it may be observed that the sound output level increases considerably when the supply voltage is raised to 16 volts. However, this is poor practice, inasmuch as the rated working voltage for the transistors will be exceeded, with the expectation of early failure. This same precaution applies to servicing intermittent conditions in amplifiers. As an illustration, an intermittent can sometimes be initated by switching the amplifier off and on, or by applying subnormal or abnormal voltage to the circuits. It is permissible to switch the power off and on for this purpose, and it is also permissible to reduce the power-supply voltage. On the other hand, it is very poor practice to attempt to trigger an intermittent condition by increasing the power-supply voltage above rated value. As noted above, this procedure is likely to damage transistors in the amplifier.

Intermittent conditions can sometimes be triggered by mechanical shock to the amplifier chassis, or by tapping various components. When this procedure is employed, it is advisable to avoid violent shocks or jarring; otherwise, damage to circuit boards with heavier components is likely to occur. When components are tapped to check for intermittents, a pencil or tuning wand should be used. Avoid the use of heavy metal tappers, such as hex-nut drivers or screwdrivers that could damage small components or disturb their connections. Of course, it is necessary to have an intermittent condition present before it can be analyzed by voltage, resistance, or waveform checks. In the case of thermal intermittents, it is often

CONTROLS (All wattages 1/2 watt, or less, unless listed)

ITEM No.	FUNCTION	RESIST- ANCE	REPLACEMENT DATA				
			MFGR. PART No.	CENTRALAB PART No.	CLAROSTAT PART No.	CTS–IRC PART No.	MALLORY PART No.
VR 501L	Tone (Left Front)	50K	0514-0080/ 6438				
VR 501R	Tone (Right Front)	50K					
VR 502LF	Volume (Left Front)	50K	0514-0099/ 6438				
VR 502RF	Volume (Right Front)	50K	0514-0099/ 6438				

RESISTORS (Power and Special)

ITEM No.	RATING		WORKMAN PART No.	MFGR. PART No.
R134	270	1W,Film		08570703811/6438
R575	3.3	1/4W,Carbon		0834-0513/6438
R576	3.3	1/4W,Carbon		0834-0513/6438

ITEM No.	RATING		WORKMAN PART No.	MFGR. PART No.
R579	3.3	1/4W,Carbon		0834-0513/6438
R580	3.3	1/4W,Carbon		0834-0513/6438
R581	3.3	1/4W,Carbon		0834-0513/6438

COILS (RF–IF)

ITEM No.	FUNCTION	REPLACEMENT DATA			
		MFGR. PART No.	OTHER IDENTIFICATION	MEISSNER PART No.	MILLER PART No.
L101	FM Antenna	0320-0091/6438			
L102	FM RF	0331-0094/6438			
L103	RF Choke	0333-0014/6438			
L104	FM Oscillator	0332-0188/6438			
L105	Loopstick	0310-0720/6438			
L201	67 kHz Trap	0356-0014/6438			
L601	Line Choke	0340-0025/6438			
L602	Line Choke	0340-0025/6438			
T101	FM Input IF	0352-0130/6438			
T102	FM Interstage IF	0352-0121/6438			
T103	FM Interstage IF	0352-0121/6438			

TRANSFORMER (Power)

ITEM No.	RATING			REPLACEMENT DATA				NOTES
	PRI.	SEC. 1	SEC. 2	MFGR. PART No.	STANCOR PART No.	THORDARSON PART No.	TRIAD PART No.	
T601	120 VAC @ 17 A AC	32.6 VAC CT @ .6 ADC	13.62 VAC @ .14 ADC Tap @ 6.1 VAC @	0201-1941/6438 AC167 (1)				(1) Number on unit.

SPEAKER

ITEM No.	TYPE			REPLACEMENT DATA		NOTES
				MFGR. PART No.	QUAM PART No.	
	6 1/2"	PM	8 ohms	69-003-001 (1)		(1) Number on unit
	6 1/2"	PM	8 ohms	69-003-001 (1)		
	6 1/2"	PM	8 ohms	69-003-001 (1)		
	6 1/2"	PM	8 ohms	69-003-001 (1)		

TAPE HEADS

ITEM No.	MEASURED			MFGR. PART No.	NORTRONICS PART No.	DESCRIPTION
	INDUCTANCE	BIAS/ERASE VOLTS (RMS)	BIAS FREQ.			
H301	490 mH					8-Track Quad Play

MISCELLANEOUS

ITEM No.	PART NAME	PART No.	NOTES
L603	Solenoid		Tape, Head Shift
M1	Motor		Tape
SW101	Switch	5017-0040/6438	Function (AM/FM/FM Stereo/Phono-Aux/Tape
SW301	Switch		Discrete
SW601	Switch		Matrix
SW601	Switch	5012-0017/6438	Power AM/FM
SW602	Switch		Tape Power (Actuated by insertion of cartridge)
SW603	Switch		Tape Power (Actuated by insertion of cartridge)
SW604	Switch		Tape Power (Actuated by insertion of cartridge)
SW605	Switch	5013-0016/6438	Manual Program Select

Figure 3–22. Typical replacement parts listing. *(Redrawn by permission of Howard W. Sams & Company, Inc.)*

effective to warm suspected components moderately, as with a hair dryer. Or, the suspected component may be sprayed with a coolant mist to reduce its temperature substantially. If the foregoing approaches do not cause the intermittent condition to appear, the amplifier should be monitored. In other words, the amplifier is operated at maximum rated output as long as may be required for the intermittent condition to appear. It is good practice in this situation to have voltmeters connected at key points through the amplifier circuit, so that the general location of the trouble is evident at the onset of the intermittent.

When defective components are replaced, it is desirable to use exact replacements. However, this is not always feasible. In such a case, the technician usually chooses an equivalent replacement. Such replacements are indicated in the amplifier service data, as exemplified in Fig. 3–22. Situations occur, however, in which service data are not available for a particular amplifier. In such a case, the technician is thrown upon his own resources in selecting suitable component replacements. This requirement can be treacherous when unidentified components such as transistors or thermistors must be contended with. Selection of replacement components with suitable ratings is dependent upon a good understanding of circuit action and component characteristics. This ability can be acquired only by persistent study and extensive practical experience.

REVIEW QUESTIONS

1. What is meant by complementary push-pull action in an amplifier?
2. Does the CE, CB, or CC mode of operation provide the most nearly equal input and output resistances in transistor amplifier operation?
3. Discuss the distinctions between voltage gain, current gain, and power gain.
4. State the purpose of a high-frequency rolloff network.
5. Explain the action of a low-frequency boost network.
6. Name one advantage of degenerative feedback in an amplifier.
7. Describe a disadvantage of objectionable circuit loading by a voltmeter.

8. State the "most likely defect" in a typical audio amplifier that employs transistors, resistors, electrolytic capacitors, paper capacitors, and mica capacitors.

9. Explain the procedure that is followed in a turn-off test; in a turn-on test.

10. Is it always possible to make a turn-off test of a transistor in any circuit?

11. Discuss the requirements for in-circuit measurement of resistor values in a solid-state amplifier.

12. Is it always possible to measure the value of any resistor in-circuit?

13. Distinguish between a depletion-type and an enhancement-type FET.

14. How can a suspected open-circuited capacitor be quick-checked in-circuit?

15. What precaution should be observed when making a capacitor "bridging" test?

16. List the numerical values indicated by the standard resistor color code.

17. Are capacitor color codes completely standardized?

18. Why is a power resistor substituted for the speaker in amplifier servicing procedures?

19. Explain the general procedure employed to signal-trace an audio amplifier.

20. Could you "see" 1% harmonic distortion in a sine wave displayed on the screen of an oscilloscope?

21. What is the most common cause of harmonic distortion in an audio amplifier?

22. Briefly describe the circuit actions that occur in an amplifier with one negative-feedback loop.

23. How do negative-feedback loops affect the evaluation of trouble areas in signal-tracing procedure?

24. Discuss the general procedure that is followed in troubleshooting intermittents.

Troubleshooting Solid-state Power Supplies

4.1 GENERAL CONSIDERATIONS

Most power supplies utilize semiconductor diodes (rectifiers) and filter circuits; this type of power supply is energized by an ac source. Figure 4–1 shows a basic half-wave rectifier and filter arrangement. Note that the load resistor R_L may be a bleeder resistor,* or R_L may represent the current demand of an amplifier, for example. Again, R_L may consist of a bleeder resistor plus the current demand of an amplifier or other unit of equipment. Troubleshooting of a power supply involves dc-voltage measurements with a TVM, ripple waveform checks with an oscilloscope, occasional resistance measurements with an ohmmeter, and occasional current measurements with a milliammeter or ammeter.

An ideal filter supplies a pure dc output, and an elaborate filter may approach the ideal. However, most filters have at least a little variation (ripple) in their output voltage. With reference to Fig. 4–1, suppose that the filter capacitor C becomes open-circuited. In such a case, the voltage across R_L consists of half-sine waves. It follows that an oscilloscope is a very useful instrument for checking

*A bleeder resistor acts to protect the circuits if a load is lost. A bleeder resistor also serves to discharge the filter capacitors in case the load becomes open-circuited. Sometimes, a bleeder resistor is used to improve the regulation of a power supply.

power-supply filter capacitors. When the output waveform has a half-sine shape, as in Fig. 4–1(b), the output is useless for energizing an amplifier. This extreme ripple voltage will cause the amplifier to develop a very loud 60-Hz sound output. Note that although the filter output voltage is varying between a peak value and zero, a dc-voltmeter check would indicate that there is a steady dc output but would be reduced from that with a good capacitor. This misleading indication is obtained because the pointer and movement of a meter cannot vibrate at a 60-Hz rate.

Next, suppose that capacitor C in Fig. 4–1 has lost a substantial portion of its rated capacitance. In such a case, partial filtering is obtained, as seen in Fig. 4–1(c); the small capacitance tends to hold a charge from one half-cycle to the next. However, the ripple waveform still varies between a peak value and zero, essentially as in (b). Accordingly, the output is useless for energizing an amplifier. Now, if the value of C is increased considerably, it holds a charge more effectively from one half-cycle to the next, as shown in (d). Although the ripple voltage is still higher than can usually be tolerated, the output variation is considerably reduced. It follows that if the value of C is very greatly increased, the voltage across the load will be practically pure dc, as depicted in (e). A ripple voltage of 10 percent or greater is usually regarded as a trouble symptom—that is, if the ripple (ac) voltage is 10 percent of the dc.

Troubleshooting tests must often be chosen to distinguish between different possible causes of a symptom. As an illustration, suppose that a waveform check at the output of the filter shows extreme ripple, as in Fig. 4–1(c). It may happen that the filter capacitor C is not defective, but that the load R_L has an abnormally low value. In other words, an excessive current demand on the power supply has the same effect upon ripple amplitude as loss of filter capacitance. If the normal current demand is known (as is often the case), a dc-current measurement will provide helpful preliminary test data. As will be explained in more detail later, many power supplies are fused, and an excessive current demand will cause the fuse to blow.

Another important possibility of trouble in the arrangement of Fig. 4–1(a) is a defective rectifier diode. A diode may develop a poor front-to-back ratio, or it may become open-circuited or short-circuited. A poor front-to-back ratio results in a subnormal dc-voltage output and increased ripple amplitude from the power supply. An open-circuited diode results in zero output voltage. A short-circuited diode causes damage in some cases to electrolytic filter capacitors, and results in ac output voltage instead of dc output voltage. In other

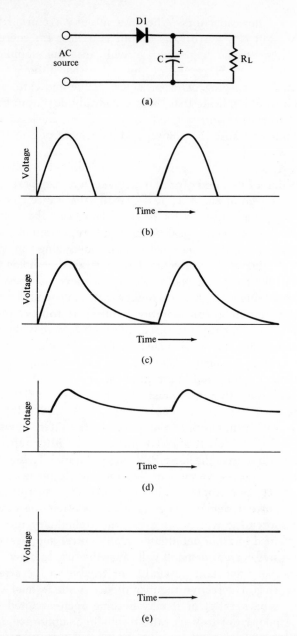

Figure 4–1. Basic half-wave rectification and filtering. **(a)** Configuration; **(b)** Load voltage with C open-circuited; **(c)** Load voltage with small value of capacitance; **(d)** Load voltage with larger value of capacitance; **(e)** Load voltage with very large value of capacitance.

words, if an oscilloscope check were made across R_L, and a sine wave were displayed, the technician would conclude that diode D1 is short-circuited.

4.2 COMMON RECTIFIER-FILTER CONFIGURATIONS

Observe next the basic full-wave rectification and filtering configuration shown in Fig. 4–2. Full-wave rectification is provided by two diodes operating in combination with a center-tapped secondary winding on power transformer T. Note that full-wave rectification produces a 120-Hz ripple frequency, instead of a 60-Hz ripple frequency as noted for the half-wave configuration in Fig. 4–1. As before, a dc voltmeter and oscilloscope are the most useful troubleshooting instruments. If the filter capacitor C becomes open-circuited, an oscilloscope will display the voltage waveform shown in Fig. 4–2(c) when the voltage across R_L is checked. If C loses a substantial portion of its rated capacitance, the pattern shown in 4–2(c) will be displayed on the oscilloscope screen.

It is instructive to note the effect of rectifier diode defects on circuit action and test-instrument indication. For example, consider the result of a poor front-to-back ratio for D1 in Fig. 4–2(a). In such a case, the successive peaks of the ripple waveform will have unequal amplitudes, as pictured in (d). Again, suppose that D1 becomes open-circuited. It is evident that the circuit action is then the same as in Fig. 4–1; the full-wave system becomes a half-wave system, and the ripple frequency is often 60 Hz. On the other hand, if D1 becomes short-circuited, there is output for a limited time only; because of the very heavy current flow imposed upon D2, either this diode will heat up and burn out, or the secondary winding on T will burn out. Of course, if the power supply is adequately fused, the fuse will blow before the diode or transformer is damaged.

If C becomes short-circuited or very leaky in the example of Fig. 4–2, a very heavy current demand is imposed upon both D1 and D2. Accordingly, the rectifier diodes will burn out soon, or the secondary winding on T will heat up and burn out. As noted previously, proper fusing of the arrangement can eliminate component damage. The methods and requirements for power-supply fusing are explained in greater detail at a later point. Note that whenever an abnormal current demand occurs internally or externally, the dc output voltage of a power supply decreases. This relation of current and voltage

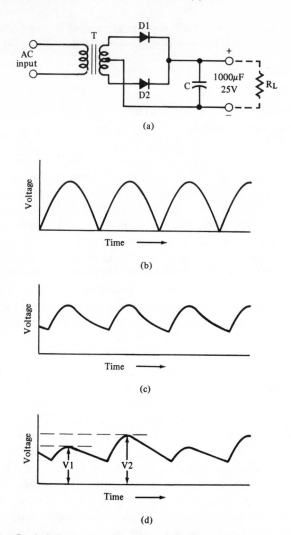

Figure 4–2. Basic full-wave rectification and filtering. **(a)** Configuration; **(b)** Load voltage with C open-circuited; **(c)** Load voltage with small value of capacitance; **(d)** Load voltage when D1 has a poor front-to-back ratio.

values is called the *regulation* of the power supply. If the voltage output does not decrease very much when the current demand is increased considerably, the power supply is said to have good regulation. On the other hand, if the voltage output falls off greatly when the current demand is increased by a small amount, the power supply is said to have poor regulation. We shall find that elaborate circuitry must be used in applications that require a power supply to keep its output practically constant as the current demand changes considerably.

Troubleshooting can be facilitated in many cases by remembering what effect filter capacitance has on the output voltage of a power supply, as measured by a dc voltmeter. These facts are depicted in Fig. 4–3. Note that when the output from a half-wave rectifier is completely filtered, the output voltage rises to the peak value, as measured with a dc voltmeter. This is the result of charging the filter to the peak value, and holding this peak value from one peak to the next. On the other hand, in case there is no filtering (filter capacitor open-circuited), the output voltage falls to 31.8 percent of peak value. In other words, the dc voltmeter now indicates the average value of the pulsating dc output.

Somewhat the same response is observed for a full-wave rectifier arrangement, as depicted in Fig. 4–3. That is, when complete filtering is provided, a dc voltmeter will indicate the peak value of the rectifier waveform. On the other hand, in case there is no filtering (filter capacitor open-circuited), the output voltage from the power supply falls to 63.7 percent of peak value. In other words, the dc voltmeter is indicating the average value of the pulsating dc output. Note that if the technician utilizes a peak-indicating ac voltmeter, he can read the peak voltage of the output waveform. This is a practical troubleshooting aid, inasmuch as the difference between the peak-voltage reading and the dc-voltage reading of the meter is equal to the ripple amplitude. Note also that if a TVM has a so-called "output" function, we can read the ripple voltage value directly, using this function of the meter.

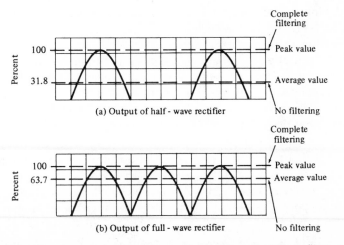

Figure 4–3. Direct current voltmeter readings with and without filter action. **(a)** Comparative readings for half-wave power supply; **(b)** Comparative readings for full-wave power supply.

In order to avoid the necessity for a center-tapped secondary winding on the transformer, some full-wave power supplies employ the full-wave bridge rectifier configuration pictured in Fig. 4–4. Many practical power supplies require the use of a surge resistor to protect the rectifier diodes against damage. In other words, when the power supply is switched on, filter capacitor C has no charge, and its initial impedance is zero. Therefore, the filter capacitor "looks like" a short-circuit to the rectifiers and transformer, until it has the time to charge. The surge-current rating of a rectifier diode is based on the maximum current flow that it can withstand for a brief time, such as 4 ms. A typical 750-mA silicon rectifier has a surge-current rating of 35 amperes. Note that the maximum voltage applied by the secondary is 180 volts. In turn, if the surge current is to be limited to 35 amperes, the surge-limiter resistor R_s should have a value of 5.2 ohms. It is very poor practice to defeat (short-circuit) the surge resistor for any reason during troubleshooting procedures.

In the example of Fig. 4–4, the normal current demand on the power supply is 300 mA. Therefore, a voltage drop of 1.56 volts would normally be measured across R_s, in accordance with Ohm's law. If the technician should find, for example, that an excessive voltage drop occurs across R_s, such as 5 volts, he would conclude either that the current demand on the power supply is excessive, or that R_s has drifted to an appreciably higher value. As noted previously, the most likely cause of excessive current demand is leakage in filter capacitor C. However, the equipment to which the power supply is connected may have a fault that imposes an excessive current demand. Note that if the power supply is temporarily disconnected from its load, there will be zero voltage drop across R_s, if the power supply is operating normally.

Semiconductor equipment occasionally requires dual-polarity outputs from a power supply, as exemplified in Fig. 4–5. In such a case, a pair of oppositely-polarized rectifier diodes are generally utilized, with a separate filter for each diode. Note the fuse that is provided on the primary side of the power transformer. A fuse is employed that has a somewhat greater rating than the maximum current demand from the line. As an illustration, if a maximum line current of 2 amperes is anticipated, the fuse would be typically rated for 3 amperes. A surge resistor is not included in this example; if a particular power transformer has sufficient internal impedance that excessive switching current cannot be applied to the rectifier diodes, a surge resistor is accordingly omitted. In some power supplies that are energized directly from the 117-volt 60-Hz line, the surge re-

Figure 4–4. Full-wave bridge rectifier and filter configuration. **(a)** Circuit with operating voltages; **(b)** Relations of rms, peak, and peak-to-peak values.

sistor is designed as a fusible resistor that will burn out if an excessive current demand is placed on the power supply. Thus, the resistor does double duty. It is very poor practice to "cheat" or defeat a fuse during troubleshooting procedures, inasmuch as costly damage may result to the power transformer and/or other components. Note that the transformer will "buzz" audibly, if it is seriously overloaded.

Observe in Fig. 4–5(b) that each of the diodes in the foregoing power-supply configuration must withstand an inverse peak voltage equal to double the value of the applied ac peak voltage. In other words, when the applied voltage swings negative, it adds in series with the filter-capacitor voltage; this double-peak voltage is blocked solely by the diode. Therefore, the diode must be suitably rated for peak inverse voltage value. This is an important point to keep in mind when one is replacing one type of rectifier diode with another type. Note also that the maximum current demand placed on rectifier diodes and fuses is sometimes substantially greater during the warm-up time

(a)

(b)

Figure 4–5. Half-wave power supply with dual-polarity outputs. **(a)** Configuration; **(b)** Peak inverse voltage applied to diodes.

of the associated equipment, than after it "settles down" to steady operation. As a simple example, if the equipment includes a small dc motor, the motor will draw several times its rated running current before it comes up to normal speed. In such situations, slow-blow fuses are utilized in the power supply instead of conventional fuses. It is essential to replace a slow-blow fuse with an equivalent unit; in other words, a conventional fuse will not "hold" during the warm-up time in this situation.

When the trouble symptom is a blown fuse, it is always necessary to determine why the fuse overheated. There are two general approaches to this problem. One method consists of placing a series resistor in the line, to limit the current demand. In turn, the new fuse will not blow, although the output voltage from the power supply is subnormal. With the power supply operating at reduced voltage, the technician may then investigate the current demands of the various output branch circuits, and localize the branch that is drawing excessive current. Then attention can be directed to pinpointing the defective component in this branch circuit. The other method consists

of temporarily disconnecting all of the load circuits from the power supply, and replacing the blown fuse. Then the load circuits are reconnected one by one. When one of the load circuits blows the fuse upon reconnection, the branch circuit that draws excessive current is thereby localized.

Sometimes a circuit breaker is provided instead of a fuse in a power supply. (See Fig. 4–6.) Although a circuit breaker can be reset, it will promptly trip in the presence of an overload. Therefore, the troubleshooting approaches are basically the same as for fused power supplies. Just as a slow-blow fuse may be provided in some power supplies, so may slow-action circuit breakers be employed. If a circuit breaker must be replaced, it is essential to replace it with an equivalent unit. Most circuit breakers utilized in small- and moderate-sized electronic equipment are of the thermal type. Large circuit breakers are likely to be of magnetic design. Although a magnetic breaker can be immediately reset, a thermal breaker cannot be reset until it has cooled down to normal operating temperature.

A dual-polarity power supply may employ a center-tapped power transformer with a full-wave bridge rectifier, as shown in Fig.

Figure 4–6. Typical fuses and circuit breakers used in small power supplies. **(a)** Manual reset type of circuit breaker; **(b)** Slow-blow-fuse; **(c)** Chassis-mounting fuseholder; **(d)** Panel-mounting fuseholder. *(Courtesy of Bussman Mfg. Div.)*

4–7. Note that the center-tapped secondary causes each half of the bridge rectifier to function as a conventional two-diode full-wave rectifier with a chosen output polarity. In other words, C1 is charged via D3 and D4, whereas C2 is charged via D1 and D2. The ripple frequency of each output is normally 120 Hz. Note that if D3 becomes open-circuited, the positive output develops a 60-Hz ripple frequency, whereas the negative output is unaffected. On the other hand, if D2 becomes open-circuited, the negative output develops a 60-Hz ripple frequency, whereas the positive output is unaffected. Full-wave bridge arrangements may utilize individual diodes, or they may employ packaged units, such as illustrated in Fig. 4–7(b). Although a packaged unit is comparatively compact and simple to install, it has the disadvantage that if one of the diodes becomes faulty, the other three diodes must be replaced also.

Various power supplies for semiconductor equipment provide dual-voltage outputs, as depicted in Fig. 4–8(a). Note that one end of

(a)

(b)

Figure 4–7. Full-wave bridge configuration with dual-polarity outputs. **(a)** Circuit; **(b)** Bridge rectifier units. *(Courtesy of International Rectifier Corp.)*

(a)

(b)

Figure 4–8. Full-wave bridge power supply with high and low voltage outputs. **(a)** Configuration; **(b)** Example of multitapped power transformer; **(c)** Appearance of transformer. *(Courtesy of Triad)*

(c)

the rectifier bridge is grounded. In turn, the circuit functions as a conventional bridge rectifier to provide a ++ output, and also functions as a two-diode full-wave rectifier to provide a + or half-voltage output also. When adjustable output voltage is required, a multitapped power transformer may be utilized, as shown in Fig. 4–8(b) and (c). This type of power supply employs a tapped primary, with two separate primary windings. These windings may be con-

nected either in series-aiding or series-opposing positions, thereby doubling the range of output voltages provided by the taps alone.

4.3 VOLTAGE-MULTIPLIER POWER SUPPLIES

Troubleshooting of voltage-multiplier power supplies involves the same general principles explained above, except that analysis of trouble symptoms is less straightforward. Basic voltage-multiplier configurations are shown in Fig. 4–9. Both half-wave arrangements are in general use. When a multiplier is energized directly from the ac line, instead of a power transformer's being used, one input line must be grounded. Two half-wave voltage doublers can be operated in series, as shown in (d), to obtain a full-wave voltage quadrupler. Similarily, a full-wave voltage sextupler can be obtained by connecting two half-wave voltage triplers in series. Note that the regulation of a quadrupler or a sextupler is poor unless the filter capacitors are quite large. If the electrolytic capacitors age to an extent that they lose a certain fraction of their rated capacitance, the regulation will degrade to a much greater extent than in ordinary power supplies.

Figure 4–9. Basic voltage-multiplier configurations. **(a)** Full-wave doubler; **(b)** Half-wave doubler; **(c)** Half-wave tripler; **(d)** Full-wave quadrupler.

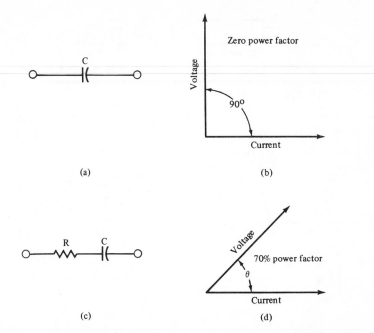

Figure 4–10. Examples of electrolytic-capacitor power factors. **(a)** Ideal capacitor; **(b)** Power factor of ideal capacitor is zero; **(c)** Capacitor with effective series resistance; **(d)** Effective resistance causes an increase in the power factor.

Leaky capacitors and diodes with deficient front-to-back ratios are also common causes of subnormal output voltage under normal load. Occasionally, an electrolytic capacitor will develop a poor power factor, although it has its full rated capacitance value, and is not leaky. This type of fault also degrades the regulation of a multiplier-type power supply. Figure 4–10 shows the equivalent circuit for an electrolytic capacitor with a poor power factor. The only practical way to check with certainty for a poor power factor is to test the capacitor on a capacitance bridge that has a power-factor scale. This test must be made out of circuit; in other words, one of the leads to the electrolytic capacitor must be opened before it can be tested with a capacitance bridge.

4 . 4 SERVICING REGULATED POWER SUPPLIES

When constant output voltage is required under conditions of varying current demand, a regulated power supply is utilized. Regulated power supplies also have less ripple voltage under load than con-

(a)

(b) (c)

Figure 4–11. A simple Zener-diode regulated power supply. **(a)** Configuration; **(b)** Zener characteristic and operating interval; **(c)** Current distribution; (left) no load; (right) half load.

ventional power supplies. As an illustration, a typical unregulated power supply may have a ripple voltage in the order of 10 percent. On the other hand, a typical regulated power supply may have a ripple voltage in the order of 0.1 percent. Thus, a regulator operates both as a voltage stabilizer and a filter. A simple Zener-diode regulated power supply is depicted in Fig. 4–11. It regulates the output voltage by virtue of its very low internal resistance in the conduction state. This is just another way of saying that the voltage drop across the Zener diode changes very little, although the current through the diode may change considerably.

Note in Fig. 4–11 that the series resistor R1 limits the current that can be drawn by the Zener diode. In turn, the output current from the power supply divides between the diode and the load R_L. If the load demands excessive current, too little current is then drawn by the Zener diode, and the operating point moves out of the normal operating interval. The result is that the output voltage becomes subnormal, and the ripple voltage becomes excessive. On the other hand, if the Zener diode becomes open-circuited, the output voltage increases

and the ripple voltage may become excessive. If the Zener diode be-comes short-circuited, the output voltage will be zero, and the pro-tective device (fuse or circuit breaker) will remove ac power from the unit. If the Zener diode develops a defect that draws excessive current short of blowing the fuse, a loud hum will be produced by the power transformer, with rapid overheating both of the transformer and of R1.

With regard to replacement Zener diodes, note that these devices are available with widely different breakdown voltage ratings, typically from 3.9 volts to several hundred volts, and with tolerances of 5, 10, and 20 percent. Power ratings range from 200 mW to 50 W, and more. The internal resistance of a Zener diode in its breakdown interval is normally less than one ohm for low-voltage types, and up to 300 ohms for the high-voltage types. Note that the internal re-sistance of a Zener diode tends to become less as its operating temper-ature increases. In case a suitable replacement happens to be un-available, the troubleshooter may connect two or more Zener diodes in series to obtain a combination with the required voltage rating, as shown in Fig. 4–12. It is only necessary that each of the diodes have a current rating at least equal to the original diode.

Some regulated power supplies employ series-connected Zener diodes with branch circuits to provide various regulated voltages, as exemplified in Fig. 4–13. If any one of the Zener diodes becomes open-circuited, all of the branch voltages are zero, although the +84-volt output continues operating at increased voltage and with a high ripple. Next, if any one of the diodes becomes short-circuited, only its associ-ated branch circuit becomes "dead." Note that the +84-volt output drops in voltage, but its ripple does not increase. For example, sup-pose that D3 becomes short-circuited. In such a case, the +51-volt

Figure 4–12. Zener diodes may be series connected to obtain increased output voltage.

Figure 4–13. Stacked Zener diodes provide multiple output voltages. **(a)** Basic configuration; **(b)** Current distribution.

output remains unaffected, and the $+84$-volt output drops to $+66$ volts. Resistor R1 will run hotter, because of the reduced effective load resistance. These facts are helpful to keep in mind during preliminary troubleshooting procedures.

As noted above, Zener diodes are rated for breakdown voltages as low as 3.9 volts. When a regulated power supply is required to provide a still lower value of output voltage, two Zener diodes are employed, as depicted in Fig. 4–14. In this example, one diode is rated at 6.8 volts, and the other diode is rated 5.6 volts. Observe that the load voltage in this circuit is equal to the difference between the two diode voltages, or 1.2 volts. This basic configuration is employed to obtain as small values of regulated supply voltage as may be needed for specialized semiconductor equipment. Troubleshooting this type of power supply is basically the same as for simpler arrangements, except that the trouble symptoms are somewhat different. In other words, if either D1 or D2 becomes short-circuited, the voltage increases greatly; the ripple remains practically unaffected. On the other hand, if either D1 or D2 becomes open-circuited, the load voltage increases greatly and the ripple becomes abnormally high.

More elaborate types of regulated power supplies utilize both Zener diodes and transistors. Their operation is based on the reference voltage established by a Zener diode, with a feedback circuit that senses any change in output voltage with respect to the reference voltage, thereby developing a control signal that is applied to a transistor to cancel the change in output. The transistor that controls the output voltage may operate in a series circuit or in a shunt circuit. Figure 4–15 shows a block diagram for a basic series voltage regulator. Note

Figure 4–14. Zener-regulated low-voltage power supply. **(a)** General plan; **(b)** First equivalent circuit; **(c)** Second equivalent circuit.

that the output proceeds through a series regulated (pass) element, which is a power-type transistor in this example. A voltage divider connected across the output terminals of the power supply feeds a sample of the output to an error detector (comparator). The difference between the sample voltage and the reference voltage is applied to the pass transistor, which in turn conducts more or less as required to cancel the change in output voltage.

It is evident that the pass transistor in a series voltage regulator conducts very little current when the current demand on the power supply is very small. On the other hand, when the current demand on

Figure 4–15. Block diagram for a basic series voltage regulator. *(Adapted from an RCA original)*

the power supply is heavy, the pass transistor conducts a large amount of current. Therefore, a series voltage regulator has high efficiency for small current demands, and least efficiency for large current demands that require large power dissipation in the pass transistor. From a practical viewpoint, the troubleshooter should keep in mind that even a momentary "dead short" on the output of a series regulator can destroy the pass transistor. In other words, it is not feasible to employ a pass transistor with a maximum current rating equal to the full short-circuit capability of the power supply. Note that a fused output circuit will guard against external overcurrent demand. However, the fuse will not protect against internal overcurrent demand caused by a component defect.

A basic configuration for a series voltage regulator is shown in Fig. 4–16. A bridge rectifier applies 25 volts dc to the series regulator transistor. The base bias voltage for the transistor is established by a Zener diode (or a choice of Zener diodes with different voltage ratings). C1 functions as a filter capacitor. C2 is a bypass capacitor; if switching transients or other transients occur in the arrangement, this potential interference is bypassed. The value of R1 is chosen to provide rated breakdown (operating) current for the Zener diode. A one-watt Zener is employed in this example. R2 is a bleeder resistor; it ensures that the regulator load will never be completely open-circuited. In other words, regulator action may become erratic for extremely small or zero current demand. Note that the output voltage of the regulator is normally equal to the Zener voltage minus the base-emitter bias voltage on Q1. Under minimum load, this bias voltage is almost zero; on the other hand, with a one-ampere load present, the bias voltage is approximately 0.3 volt.

The principal circuit action of this series voltage-regulator arrangement is shown in the equivalent circuit of Fig. 7–16(b). From a practical viewpoint, the Zener diode operates as a base-bias battery for the transistor. When the power supply is turned off, no current flows through R_L. Next, when the power supply is turned on, Q1 starts to conduct heavily, because its base is forward-biased and its emitter voltage is zero. However, as soon as the output current rises to the point that the IR drop across R_L is almost −6 volts, Q1 approaches cutoff. However, Q1 will not cut off completely, unless R_L is open-circuited. Note that if R_L were accidently short-circuited, the emitter voltage on Q1 could not rise above zero, no matter how much current flows. Therefore, Q1 would continue to operate with 6 volts of forward bias and would promptly burn out.

When Q1 is replaced during servicing procedures, it must be

(a)

(b)

Figure 4–16. Basic configuration for a series voltage regulator. **(a)** Detailed circuit; **(b)** Basic equivalent circuit. *(Adapted from an RCA original)*

carefully secured to its heat sink, which should be thermally coupled to the main chassis of the power supply. Since the transistor must be electrically insulated from its heat sink, a mica spacer is employed. This spacer should be coated with silicone grease to insure adequate thermal conductivity. If a trouble symptom of abnormal ripple occurs, C1 is the prime suspect. In other words, the regulator has good filtering action, provided that the input ripple amplitude does not exceed a certain limit. On the other hand, with excessive input ripple amplitude present, the regulator action will become more or less erratic. In such a case, the output ripple amplitude increases, and the output voltage may be subnormal under appreciable load. The trouble symptoms are typically intensified when the load current is increased.

Next, a more elaborate series voltage regulator arrangement

is shown in Fig. 4–17. Note that the output voltage decreases from 49 to 48 volts over a current-demand range from 0 to 400 mA. The percentage regulation is 2 percent up to 350 mA, and the peak-to-peak output ripple is less than 0.3 volt at 400 mA load current. At no load, the ripple amplitude decreases to 0.01 volt p-p. In other words, the ripple voltage is approximately 0.6 percent of the output voltage at a 400-mA current demand. This regulator configuration has a low output impedance; at current demands less than 200 mA, its output impedance is somewhat less than 2 ohms, and at current demands greater than 200 mA, its output impedance is less than 1 ohm.

With reference to Fig. 4–17(a), Q1 is the series regulator transistor, and Q2 is a feedback amplifier transistor. Zener diode D1 establishes a reference voltage of 22 volts, and is energized via R3.

(a)

(b)

Figure 4–17. A more elaborate series voltage regulator arrangement. **(a)** Configuration; **(b)** Regulation characteristic. *(Courtesy of General Electric Semiconductor Products Dept.)*

Note that the base bias on Q2 is determined by a voltage divider connected to the output bus of the power supply. Since the emitter voltage on Q2 is established by the Zener diode, its base-emitter voltage effectively senses any change in output voltage. This change is amplified by Q2 and is applied to the base of Q1. In turn, the collector current of Q1 changes in the required amount and direction to maintain the output voltage essentially constant. It is helpful to compare this arrangement with the configuration shown in Fig. 4–16. Observe that the same fundamental regulating principle is employed, except that amplified control action is utilized in the circuitry of Fig. 4–17. Amplified control action results in an improved regulation characteristic.

Troubleshooting is required in case the regulated power supply develops abnormal ripple and/or an impaired regulation characteristic. In some cases, the failure might be catastrophic, with a "dead" trouble symptom, although the fuse is not blown. This situation can result from an open-circuited control transistor Q2, with resulting excessive output voltage and associated current demand. In turn, Q1 probably burns out, and the stored energy in the filter is sometimes sufficient to open-circuit Q1 before the fuse blows. Note however, that trouble symptoms are more likely to be caused by defective electrolytic capacitors than by defective semiconductor devices. Therefore, in case of doubt, always check the electrolytic capacitors first. Trouble analysis is sometimes inconclusive, owing to circuit interactions. Accordingly, the technician may have to proceed with a systematic checkout of each component in turn. After component replacement, the 5-k bias potentiometer in Fig. 4–17 may need touch-up adjustment to obtain the correct value of output voltage with optimum regulation.

Next, it is instructive to consider a regulator circuit with a comparatively wide range of output voltage. For example, the arrangement depicted in Fig. 4–18 provides an adjustable output voltage from 5 to 15 volts, at a maximum current demand of 5 amperes. A maximum source voltage of 50 volts may be applied to this regulator configuration. If an output voltage less than 5 volts or greater than 15 volts is desired, the values of R2, R3, and R4 are changed in accordance with the values in Table 4–1. When suitable values are used, and R3 is properly adjusted, output voltages from 2 to 40 volts can be obtained. Note that the value of R1 determines the shutdown current (maximum current that will flow with the output terminals short-circuited). Hence, the value of R1 is chosen to protect the regulator transistor or the power transformer, depending on which component has the smaller current rating.

Since an integrated circuit is used in the configuration of Fig.

4–18, the arrangement has the appearance of simplicity. However, if an IC were not utilized, a large number of discrete components would be required and the circuit would appear complicated. Troubleshooting the IC regulator unit is comparatively straightforward. In the case of a "dead" regulator trouble symptom, the pass transistor falls under suspicion first. A voltmeter check should be made to verify the assumption that the source voltage is normal, and that the output stops at the power transistor. If the source voltage is very poorly filtered (has an excessively high ripple amplitude), the action of the regulator may be impaired. Therefore, an oscilloscope should be used to check the input ripple level to the regulator before it is assumed that erratic operation is being caused by a defective regulator component.

Next, it is instructive to consider the basic shunt voltage regulator configuration shown in Fig. 4–19. A shunt regulator has comparatively poor efficiency, but has the advantage of inherent overload and short-circuit protection. Note that the basic shunt regulator has a shunt element that determines current distribution, and a reference-voltage element. If the load current or input voltage changes, the change is reflected as a voltage variation across R1, which is connected in series with the load. In turn, the shunt-element current flow is modified so that the output voltage remains essentially constant. Observe that the power transistor is connected in the CE configuration across the load. In this example, two power transistors are shown, connected in a Darlington circuit. A Darlington configuration is basically similar to a parallel arrangement of the transistors, but has the advantage of much better control action, owing to the high current gain that is provided.

In the example of Fig. 4–19, a simple forward-biased diodes is

Figure 4–18. A series voltage regulator circuit with adjustable output voltage. *(Adapted from an RCA original)*

TABLE 4-1

Resistive Values for Various
Voltage and Current Outputs
(Figure 4-18)

Voltage Divider			Current Limit	
V_{OUT}	R_A*	R_B*	I_{MAX}	R_1
3.6	6135	2967	.05	12
5	4417	3654	0.1	6
9	11,043	2442	0.5	1.2
12	14,724	2314	1.0	0.6
13.6	16,687	2272	1.5	0.4
15	18,405	2243	2	0.3
20	24,540	2177	2.5	0.24
28	34,356	2122	3.0	0.20
			5	0.12
			10	.006

*R_A – R2 plus top portion of R3.
R_B – R4 plus bottom portion of R3.

employed as a voltage reference for the Darlington configuration. When more precise regulation is required, a Zener diode is utilized instead of a conventional junction diode. Observe that if the output voltage tends to decrease, the load current also tends to decrease. Since part of the load current branches through D1, less current flows out of the emitter of Q1 and into the base of Q2. Since Q2 then conducts less collector current, more of the incoming current is diverted from R1 into the load, and the output voltage tends to increase accordingly. A shunt regulator arrangement has good efficiency only when it is operated near its maximum rated output current. Note that a Darlington shunt element provides good regulation because of its high current gain, compared with that of a single transistor.

It is evident that if the output terminals are short-circuited in Fig. 4–19, no damage results to the transistors or diode, because all of the available current then flows through the load, and none flows through the semiconductor devices. Good regulating action depends upon a suitable value for R2, and if the Darlington pair is replaced, it may be necessary to change the value of R2 somewhat, as R2 sets the bias current for Q1. R1 serves as a current-limiting resistor, and its value is chosen sufficiently high that the power transformer will not be damaged in case R_L becomes short-circuited. Note that the unregulated input must be reasonably well-filtered, although the regulating circuit has a high degree of filtering action in normal operation. If diode D1 becomes open-circuited, the output voltage and current rise to maximum values, and there is no regulating action.

(a)

(b)

Figure 4–19. Basic Darlington-type shunt voltage regulator. **(a)** Block diagram; **(b)** Basic regulator circuitry. *(Adapted from an RCA original)*

REVIEW QUESTIONS

1. Define the term "ripple".
2. Would a 10% ripple voltage be regarded as a trouble symptom?
3. What is the ripple frequency for a half-wave power supply?
4. What is the ripple frequency for a full-wave power supply?
5. Why might successive peaks in a ripple waveform have unequal amplitudes?
6. Does an open-circuited diode in a full-wave power supply cause the circuit to operate as a half-wave supply?
7. Define the term "regulation."
8. State the average value of the output voltage from a half-wave rectifier.

9. State the average value of the output voltage from a full-wave rectifier.

10. Can ripple voltage be measured approximately on the Output function of a TVM?

11. Describe a full-wave bridge rectifier configuration.

12. Explain the origin of surge-current demand on a rectifier diode when a power supply is switched on.

13. How does a technician analyze an overcurrent trouble symptom in a power supply and its load?

14. Discuss the general arrangement of a dual-polarity power supply.

15. If a power transformer produces a loud buzzing sound, what fault is indicated?

16. Define the term "peak inverse voltage".

17. Name the two general classes of circuit breakers.

18. Describe the basic features of a voltage-multiplier power supply.

19. Is a rectifier diode or an electrolytic capacitor more likely to cause malfunction in a power supply?

20. Explain the chief characteristics of a regulated power supply.

21. How are Zener diodes connected when a regulated output of less than 3.9 volts is required?

22. Name the two general classes of Zener-diode/transistor regulated power supplies.

23. Briefly discuss amplified control action in a regulated power supply.

24. Explain the Darlington transistor configuration.

Troubleshooting Solid-state Control Circuits

5.1 GENERAL CONSIDERATIONS

Control circuits can be grouped into various categories such as motor speed controls, electronic timers, time-delay units, lamp dimmers, light-operated switches, synchronous switches, flashers, heat controls, and so on. Some control circuits process considerable power, whereas others vary or switch minute amounts of power. Semiconductors employed in control circuits include unijunction transistors, silicon controlled rectifiers, diacs, triacs, photocells, photodiodes, and light-dependent resistors, in addition to the various types of diodes and transistors discussed in foregoing chapters. Figure 5–1 shows the appearance of a unijunction transistor, with its symbol and indication of the principal associated voltages and currents.

A unijunction transistor (UJT) is a three-terminal device which has electrical characteristics quite different from those of a bipolar or unipolar transistor. Its basic characteristics are (1) a stable triggering voltage that is a fixed fraction of the applied interbase voltage V_{BB} indicated in Fig. 5–1, (2) a very low value of firing current, (3) a negative-resistance characteristic, and (4) a high pulse-current capability. The interbase resistance between B_1 and B_2 is in the range from 5K to 10K ohms. Note that B_1 and B_2 are called base-one and base-two, respectively. Next, emitter E is a rectifying contact between B_1 and B_2. In usual applications, B_1 is grounded and a positive bias voltage V_{BB} is applied to B_2. When the emitter becomes forward-

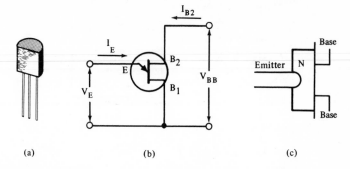

Figure 5–1. Unijunction transistor. **(a)** Appearance *(Courtesy of Raytheon)*; **(b)** Symbol; **(c)** Interior structure.

biased, a negative-resistance interval suddenly occurs in which the emitter current increases as the emitter voltage decreases. The UJT is often utilized in oscillators, timing circuits, and so on.

A silicon controlled rectifier is a three-terminal device that is normally an open circuit until an appropriate gate signal is applied to the gate terminal, at which time it rapidly switches to a conducting state. Its operation is comparable to that of a thyratron tube. Figure 5–2 shows the appearance of typical silicon controlled rectifiers, with their standard symbol and terminal designations. A silicon controlled rectifier (SCR) may be regarded as a modified rectifier, which operates as an electric switch with no moving parts. In its "off" state, an SCR blocks current flow in either direction between cathode and anode. However, if a trigger voltage is applied to its gate terminal, an SCR "turns on" instantly and conducts electrons from cathode to anode. In other words, it operates as a rectifier in its conducting state. An SCR remains turned on after the trigger voltage is removed. It

Figure 5–2. Silicon controlled rectifier. **(a)** Appearance of typical units; **(b)** Symbol; **(c)** Plan of interior structure.

(a) (b)

Figure 5–3. Diac, or bidirectional switch. **(a)** Symbol; **(b)** Interior structure.

stops conducting only when the anode voltage is removed, or when the anode voltage changes from positive to negative.

A diac is a two-terminal device that is normally an open circuit until a specific voltage of either polarity is applied. At this point the diac starts conducting; hence, the device can be regarded as an ac switch. A diac has the same general appearance as an SCR, except that it has only two terminals. Figure 5–3 shows the standard symbol for a diac, and the internal structure of the device. It returns to a nonconducting state each time that the applied voltage goes to zero; it then remains in this state until a specific voltage of either polarity is again applied, whereupon the diac again goes into sudden conduction. A triac is a three-terminal ac switch that has the same characteristics as an SCR, except that it will conduct in either direction. Its external appearance is the same as an SCR; its internal structure and its standard symbol are shown in Fig. 5–4. Like SCR's, diacs and triacs are useful in circuits that control substantial current flow and which operate at high efficiency.

(a) (b) (c)

*MT denotes mounting terminal

Figure 5–4. Triac, or three-terminal bidirectional switch. **(a)** Appearance; **(b)** Symbol; **(c)** Interior structure.

(a)

(b)

Figure 5–5. Typical photocells. *(Courtesy of Edmund Scientific Company)*

A photocell (or photovoltaic cell) is a semiconductor device that converts light into electricity. Typical selenium and silicon photocells are illustrated in Fig. 5–5. A typical output of 0.5 volt is provided by a high-efficiency silicon photocell with a maximum current output of 40 mA, when the device is exposed to bright sunlight. Note that the so-called cadmium-sulfide photoresistor is not a voltage-generating device: as mentioned in the first chapter, this device is a light-de-

pendent resistor that changes its resistance value in response to light energy.

5.2 MOTOR SPEED CONTROL

A motor speed control unit for operation of ac motors rated up to 8 amperes typically employs a diac and triac configuration, as depicted in Fig. 5–6. Potentiometer R1 provides manual control of speed. This arrangement is called a *double time-constant network*; it provides the necessary delay to trigger the triac at very low conduction angles. In turn, practically full power is provided to the load at the minimum resistance setting of R1. It is instructive to observe the basic circuit for a diac/triac phase-control circuit shown in Fig. 5–6(b). Triac Y1 is gate-controlled by diac Y2. Observe that the voltage applied to Y2 is partially out of phase with the voltage across Y1. This phase difference is determined by the value of C and also by the value of R2. In turn, the amount of current that flows through Y1 and R_L depends upon the extent to which the gate trigger voltage of Y1 is in phase with the voltage across MT1 and MT2.

(a)

(b)

Figure 5–6. Configuration of a motor speed control unit. **(a)** Detailed circuit; **(b)** Basic circuit.

Figure 5–7. Current flow in a basic motor speed control circuit.

Normal circuit action is shown by the graph in Fig. 5–7, as the value of C is varied. Note that the control action is continuously variable between minimum and maximum current values. Practically instantaneous response is obtained to a change in value of C, or of R2. Next, with reference to Fig. 5–6(a), the voltage across Y2 is adjusted by potentiometer R1. Resistor R2 is included to limit the surge across Y1, and thereby maintain operation within device ratings. Note that R3 and C2 comprise a snubber circuit; it reduces the peak voltage of transients, thereby protecting both Y1 and the control arrangement. Figure 5–8 shows the appearance of an engineering prototype motor speed control unit.

Figure 5–8. Motor speed control unit. *(Courtesy of International Rectifier Corp.)*

Troubleshooting malfunctions in the system starts with localization of the fault. In other words, either the motor or the control unit may become defective. Therefore, it is advisable first to connect a short-circuit bus across R1 and C1 in Figure 5–6. Then, if the motor develops rated power output, the trouble will be found in the control unit. Otherwise, the motor should be repaired or replaced before tests are made in the control unit. If the control unit requires troubleshooting, the possibility of related breakdowns should be kept in mind. As an illustration, if Y1 is found to be burned out, it is likely that C2 will also be open-circuited. Therefore, the technician should check C2 when replacing Y1; otherwise, the replacement triac is likely to fail in a short time. If C1 or C3 becomes leaky, the control range will be disturbed; serious leakage will prevent Y2 from conducting at all. An open circuit in C3 will reduce the normal control range. Note that potentiometers tend to become worn and "noisy" after extended use. Accordingly, in case of erratic control action R1 is the prime suspect.

5.3 ELECTRONIC TIMER

A timer that switches a circuit off after a predetermined time interval employs a MOS field-effect transistor, as depicted in Fig. 5–9. Potentiometer R3 provides adjustment of the timing interval, with a maximum period of 11 seconds for the component values illustrated. Figure 5–10 shows the appearance of a circuit board corresponding to the schematic diagram in Fig. 5–9. It is instructive to consider

Figure 5–9. Configuration of an electronic timer.

Figure 5–10. Circuit board used in the electronic timer. *(Courtesy of RCA)*

the circuit action of this electronic timer. When switch S1 is thrown, S1a is opened and S1b is closed. In turn, capacitors C1 and C2 charge through rectifiers CR1 and CR2, respectively. Thereby, the voltage across C2 is applied to the gate of triac Q4 via R6 and R13, causing Q4 to turn on. Note that the voltage across C2 is also applied to the RC time-constant circuit comprising R8 and C4. As noted above, the voltage across R8 and C4 (or the time constant) depends upon the setting of R3. Note in Fig. 5–9 that C3 filters the voltage across R8 and C4. When the voltage across C4 becomes sufficiently high, Q1 starts to conduct and the MOS transistor turns on. This places a forward bias on the base-emitter junction of Q2. Consequently, Q2 turns on and triggers the regenerative switch comprising Q2 and Q3. Note that when the regenerative switch conducts, it bypasses the gate of triac Q4. In turn, Q4 turns off and the power in the load stops. The timer circuit then remains in this state until switch S1 is opened, at least momentarily. S1 is a DPDT switch that is wired so that the C4 discharge circuit is completed when the power-line circuit is opened. In turn, C4 is enabled to discharge rapidly through R9 and permits the timer to be recycled quickly.

Typical trouble symptoms include incorrect timing, "dead" timer, erratic operation, and intermittent operation. Incorrect timing may consist in abnormal delay in turnoff of Q4 (Fig. 5–9). In this situation, it is good practice to check the power supply first, both for

dc output voltage (approximately 160 volts) and for ripple amplitude. Rectifier diodes and/or filter capacitors should be replaced, as required. With the power supply operating normally, C4 falls under suspicion when turnoff of Q4 is abnormally delayed. In other words, leakage in C4 drains away part of the current from R8, and slows down charge buildup in C4. Another possible cause of this trouble symptom is extensive wear in R3, causing the charging circuit to have an abnormally high resistance. However, this is the least likely cause of trouble.

Next, in case incorrect timing consists in premature turnoff of Q4 (Fig. 5–9), R3 falls under suspicion first. For example, a worn spot near the lower end of the resistance element, or a poor contact between the element and terminal 1 will result in the application of maximum voltage to C4, with little or no control provided by R3. Although less likely, the same trouble symptom can be caused by an open-circuit in R4, or a break in the circuit-board conductor between terminal 1 and R4. Of course, a cold-soldered joint at the connection of R4 and R3 will have the same effect. Still another possibility for the same trouble symptom is a substantial drift in the value of R8 to a lower resistance. As a practical troubleshooting note, composition resistors are more likely to increase in value with age, than to decrease in value. It is evident that if C4 becomes open-circuited, Q4 will turn off almost instantaneously, with an apparent trouble symptom of "dead" timer.

Other probable causes of a "dead" timer trouble symptom in Fig. 5–9 include power-supply failure and transistor failure. To check for defective transistors, measure the collector (or drain) voltage as the timer is cycled. A good transistor will show a substantial change in voltage as it switches off and on. However, a defective transistor will show little or no change in voltage during this test. Note that if all three transistors respond normally, it would be concluded that triac Q4 is defective. When replacement transistors are soldered into the circuit board, it is essential to avoid the use of a soldering iron that has excessive ac leakage. Otherwise, the leakage current is likely to damage the new transistor. If a doubtful soldering iron must be used, be sure to run a ground wire from its metal structure to the common bus on the circuit board. This "trick of the trade" insures that the soldering iron remains at reference ground potential.

Erratic operation of the timer can be caused by loose connections or a badly worn timing control. Intermittent operation can be caused by the same defects, and also by thermally responsive capacitors or transistors. Ordinary resistors occasionally become inter-

mittent. It is more difficult to troubleshoot an intermittent condition in a timer than in an amplifier, because a timer does not operate continuously. For example, the configuration in Fig. 5–9 operates for a maximum time interval of 11 seconds. However, this is sometimes sufficient to locate a defective component or faulty connection by tapping. As noted previously, an intermittent condition can be caused by a microscopic break in a circuit-board conductor. In such a case, the intermittent can often be made to appear or disappear by moderate flexing of the board.

5.4 LAMP DIMMER CONTROL UNIT

Lamp dimmers often utilize triacs for a wide range of load current. As an illustration, Fig. 5–11 shows the appearance of a typical lamp-dimmer unit; it provides continuous control of loads up to 700 watts. A comparatively simple configuration with a single triac is employed, as depicted in Fig. 5–12. This unit supplies a range of zero to 6 amperes for control of incandescent lamps rated for 120-volt operation. As noted previously, a triac conducts in either direction, so that full-wave control of the load is provided. Note that the triac is triggered into condition on each half-cycle by a pulse applied to its gate terminal. In turn, the power supplied to the load is a function of the time over each half-cycle at which the triac is triggered into conduction, as illustrated in Fig. 5–13.

Note that the point at which the triac is triggered into con-

Figure 5–11. Appearance of a lamp-dimmer control unit. *(Courtesy of RCA)*

Figure 5–12. Configuration of a lamp-dimmer arrangement.

duction on each half-cycle is determined by the values of C1, C2, R2, and potentiometer R1 in Fig. 5–12. When the foregoing capacitors charge to the triggering voltage of neon lamp I1, the gate of the triac draws current. Then the triac conducts for the remainder of the half-cycle. As the ac supply voltage falls to zero, the triac becomes non-conducting and does not again start conduction until the neon lamp "fires" on the next half-cycle. It is important to avoid exceeding the current rating of the triac; if it is overloaded, the device will be damaged. It is also essential to keep the heat sink on the triac while one is testing or troubleshooting the unit.

It is evident from the waveform shown in Fig. 5–13 that the triac goes into conduction abruptly at the instant that it is triggered by the neon bulb. This abrupt conduction can cause radio interference; the RF voltages are radiated by the power line from the lamp dimmer. To minimize radio interference, the technician may insert a line filter, as depicted in Fig. 5–14. Coil L consists of 18 feet of No. 18 insulated wire random-wound on a wood or plastic rod ½ inch in diameter. Capacitor C is a 0.05 µf paper type rated for 200 volts. If a "dead" trouble symptom occurs, the most likely cause is leakage in C2. This fault can prevent Q1 from going into conduction although R1 is set

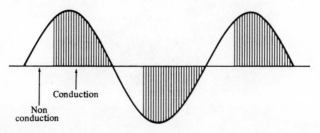

Figure 5–13. Example of triac conduction over part of each half-cycle.

Figure 5–14. An LC line filter.

to its minimum resistance position. This same trouble symptom can be caused by burnout of R1 owing to a short circuit or excessive leakage in C1.

Alternating current voltage measurements through the circuit are very helpful in preliminary analysis of a trouble symptom. Suppose, for example, that substantial voltage is measured across C2, but I1 does not glow. It is logical to conclude that the neon lamp is defective, or that the gate of Q1 is open-circuited. A useful quick check consists in temporarily short-circuiting the gate terminal of Q1 to terminal MT2. Then, if the neon lamp glows, the triac is defective; if the neon lamp does not glow, the lamp is defective. Although neon lamps are comparatively long-lived, the trigger voltage value tends to increase as the lamp ages. In turn, a point is eventually reached at which the lamp will not trigger in response to the available voltage. This change is associated with development of incorrect neon gas pressure.

5 . 5 OFF-ON HEAT CONTROL WITH SYNCHRONOUS SWITCHING

Triac heat controls are used for energizing resistance-heating elements. Figure 5–15 shows a typical configuration that provides synchronous switching near the beginning of the zero-voltage crossing of the input voltage. This method is advantageous for reduction of radio-frequency interference. Note that a thermistor R6 controls the operation of a two-transistor regenerative switch. This switch, in turn, controls the operation of the triac. If the temperature to which the thermistor is exposed becomes low, the resistance of the thermistor increases, and the regenerative switch turns off. This causes the triac to trigger directly from the ac line on positive half-cycles of the input voltage. Accordingly, nearly the full ac input voltage is applied across the heater load, and capacitor C1 charges to practically the peak value of the input ac voltage.

Next, during the negative half-cycle of the input ac voltage, C1 discharges through resistor R1, diode D1, and the triac gate.

Figure 5–15. Configuration of a typical heat-control unit.

Thereby, the triac is triggered into conduction at the start of the negative half-cycle. Note that the slaving network comprising C1, R1, and D1 insures that the triac is triggered on negative half-cycles of the input ac voltage after it has been triggered on the positive half-cycles, in order to supply integral cycles of ac power to the heater load. On the other hand, when the temperature rises to a specified value, the resistance of thermistor R6 decreases to a point that the regenerative switch is turned on. This switch then conducts at the beginning of the positive alternation of the input-voltage cycle, and shunts trigger current away from the gate of the triac. In other words, the triac does not conduct so long as the resistance of the thermistor is sufficently low that the regenerative switch turns on before the triac can be triggered.

Troubleshooting the arrangement depicted in Fig. 5–15 starts with analysis of the trouble symptoms. As an illustration, the unit may be "dead," or the triac may conduct constantly, regardless of temperature. In the first condition, the question arises concerning whether the triac is defective. Therefore, a check should be made for positive voltage at the gate of triac Y. If substantial voltage is present at the

gate terminal, it is logical to conclude that the triac is defective. On the other hand, if the gate voltage is zero, the triac is probably normal; diodes D1, D2, and D3 should be tested. If D3 is open-circuited, no trigger voltage can be applied to the gate of the triac. Similarly, in case both D1 and D2 are short-circuited, no trigger voltage can be applied. Note, however, that if the diodes are not defective, it is highly probable that the triac has failed and that there is an internal "short" between the gate and MT1.

Next, consider the situation in which a small positive voltage is measured at the gate of the triac. A subnormal trigger voltage, of course, will prevent conduction. In this situation, D3 may be found to have a high forward resistance, or there may be a defective component in the regenerative-switch circuit. As an illustration, if the thermistor has an abnormally low resistance, the regenerative switch will remain on continuously, regardless of temperature. In turn, the switch will bypass most of the trigger voltage around D3. Because of the internal resistance of the switch circuit, the gate voltage of the triac will not be zero, but will read a small positive value.

Again, suppose that the gate voltage of the triac remains above the trigger level, regardless of temperature. In such a case, the triac conducts continuously and there is no control action. Suspicion then falls upon C2 or C3. In other words, if C2 is short-circuited, Q2 remains cut off continuously. Similarly, if C3 is short-circuited, Q1 remains cut off continuously. In either case, the regenerative switch remains "open," regardless of the thermistor resistance. However, if both capacitors are in good condition, it will be concluded that Q1 or Q2 is defective owing to an internal open circuit. As noted previously, defective capacitors are more likely to cause trouble in these situations than are transistors. Rarely, an off-value resistor will cause trouble symptoms, although resistors are least likely to become defective.

5.6 LIGHT-OPERATED SWITCH

Many types of light-operated switches are in use. More recent designs often employ transistor circuitry such as that shown in Fig. 5–16. An electromechanical relay is utilized to switch the load current off or on. This is called a positive-action arrangement, meaning that the load is receiving either no power or full power. In other words, the load current does not change with the amount of light entering the photocell. Observe that the photocell energizes a regenerative switch comprising

transistors Q1 and Q2. The switch has a high internal resistance until a threshold voltage value is applied by the photocell. Thereupon, the switch is triggered and it suddenly develops a very low internal resistance. The trigger level is determined by the setting of R4. Figure 5–17 shows a circuit board for this light-operated switch.

When light enters the photocell, its resistance decreases and permits increased current flow into the base of Q2. At some point, Q2 starts to conduct, and its collector becomes less positive. This drop forward-biases Q1 into conduction. In turn, its collector becomes less negative (more positive), and Q2 conducts still more. In other words, a positive-feedback action takes place in the regenerative switch, which makes it change state with extreme rapidity. While the switch is conducting, it shunts current around relay K1, and the relay contacts open. Next, when light stops entering the photocell, its resistance increases and the positive voltage applied to the base of Q2 is then insufficient to maintain conduction in the transistor. Therefore, Q2 is cut off in turn owing to the lack of regenerative action. With Q2 cut off, the current through the relay winding increases and the relay contacts close.

Troubleshooting the circuitry shown in Fig. 5–16 is not difficult. Common trouble symptoms are "stuck at" on, or "stuck at" off. In other words, the arrangement does not respond to the presence or absence of light at the photocell, and the relay remains continuously open or continuously closed. Note that collector-junction leakage in

Figure 5–16. Configuration of a positive-action light-operated switch.

Figure 5–17. Circuit board for light-operated switch.

Q2 will disturb the threshold of conduction, and appreciable leakage will cause Q2 to conduct continuously, whether light is entering the photocell or not. In turn, the relay becomes "stuck at" off. On the other hand, in case Q2 burns out or develops an internal open circuit, no current can flow through the transistor, and the relay becomes "stuck at" on. Similarly, defects in Q2 can result in no response from the unit. If one of the transistors falls under suspicion, a substitution test is advisable because dc voltage measurements are somewhat inconclusive.

Photocells generally have higher reliability than do transistors. However, a photocell can develop changed characteristics after long service. The resistance of the photocell in Fig. 5–16 can not be measured in circuit, because it is shunted by circuit resistance. Therefore, one terminal must be disconnected for test. If the photocell is in good condition, its resistance will be less than 100 ohms when it is exposed to a bright light. Its dark resistance is normally extremely high, ranging into many megohms. Another trouble possibility is a faulty diode, although this is not highly probable. If CR1 develops a big forward resistance or becomes open-circuited, the supply voltage will be affected accordingly. If CR2 has the same fault, the relay will become inoperative although the supply voltage is unaffected.

Note that C1 in Fig. 5–16 operates as a filter capacitor. If this capacitor becomes open-circuited, the relay will tend to chatter.

On the other hand, if C1 becomes short-circuited, the relay will become inoperative. As a practical application note, the relay terminals are shown connected so that the power to the load is turned off when the light into the photocell reaches a threshold value. If it is desired to reverse the operation of the relay, it is necessary only to connect the lead from CR1 to terminal A of the relay, instead of terminal B, as shown in the diagram. The amount of current that can be controlled depends upon the rating of the relay. It is poor practice to overload a relay, because its contacts will be burned and the relay will become inoperative.

REVIEW QUESTIONS

1. Distinguish between an SCR, a diac, and a triac.
2. How does a basic motor speed control unit operate?
3. Explain the function of a snubber circuit.
4. What components are most likely to become defective in a motor speed control unit?
5. Could a conventional diac and triac control the speed of an ac motor rated for 8-ampere operation?
6. Describe the basic features of an electronic timer.
7. Discuss probable faults that would cause incorrect timing, no response, erratic operation, and intermittent operation.
8. Would it be practical to set an MOS electronic timer for a period of 11 seconds?
9. Name a suitable semiconductor device for the operation of a lamp dimmer control unit.
10. State a probable cause of RF interference from a triac-type lamp-dimmer control unit.
11. How is RF interference prevented in the foregoing situation?
12. Why does a neon lamp eventually become inoperative?
13. What kind of semiconductor device would be used in an off-on heat control with synchronous switching?
14. Describe the operation of a two-transistor regenerative switch.
15. How would a triac be tested for defective operation?
16. Explain the operating characteristic of a thermistor.
17. Define a postive-action arrangement with respect to relay circuitry.

18. Discuss the operating characteristic of a photocell.
19. What is the meaning of a "stuck at" trouble symptom?
20. State a typical resistance-variation range for a photocell from darkness to very bright illumination.
21. Identify the terminals of a diac; of a triac.
22. Explain the operating characteristics of a unijunction transistor.
23. Name two semiconductor substances utilized in fabrication of photovoltaic cells.
24. Could a silicon photocell develop an output of 0.5 volt at 40 mA?

Troubleshooting Solid-state Test Equipment

6.1 GENERAL CONSIDERATIONS

A transistor voltmeter (TVM) is similar in functions to a volt-ohm-milliammeter (VOM). However, owing to its internal transistor amplifier, a TVM such as that illustrated in Fig. 6–1 is much more sensitive. This is an example of a field-effect transistor multimeter that has very high input impedance. Other TVM's employ bipolar transistors; when comparatively elaborate circuitry is utilized, such TVM's can also provide very high input impedance. FET meters may use either JFETS's or MOSFET's. In any case, we shall find that the basic arrangement of the meter consists of a bridge circuit, two arms of which include transistors. In its resting state, the meter is balanced. On the other hand, when an input voltage is applied to the TVM, one arm of the bridge is unbalanced and a proportional current flows through the meter. In turn, the meter scale is calibrated to indicate dc voltage values. Alternating-current voltage values are indicated by passing the applied ac through an instrument rectifier before the resulting dc voltage is applied to the bridge. Resistance values are indicated by applying the voltage output from a conventional ohmmeter circuit to the bridge.

6.2 BASIC TVM CIRCUITRY

A simplified schematic diagram for the dc-voltage function of a TVM is shown in Fig. 6–2. The dc-voltage ranges are from 1 to 1000

Figure 6–1. Appearance of a transistor multimeter. *(Courtesy of Sencore)*

volts, with an input resistance of 15 megohms on all ranges. Note that a pair of field-effect transistors are used in a balanced bridge circuit. In turn, there is no current flow through the meter movement unless the bridge is unbalanced by application of an input voltage. Neon lamp NE2 is provided to protect Q1 against overload. In other words, if an excessive input voltage is applied, the neon lamp fires and bypasses the input voltage around the FET. Note that diode D1 is not a rectifier; it is employed for temperature compensation of transistor gain.

Bridge balance in Fig. 6–2 is set precisely by adjustment of the 50k zero-adjust control. This control sometimes requires a slight readjustment when dc-voltage ranges are switched. Note that the dc-balance control is a maintenance adjustment, which is set to provide a suitable operating point for the zero-adjust control. The 5 kilohms dc calibration control is also a maintenance adjustment. It requires attention only if a component should be replaced in the meter circuit. This calibration control is set to provide correct full-scale indication.

Next, consider the ac-voltage function circuitry depicted in Fig. 6–3. The ac-voltage ranges are from 1 to 1000 volts, with an input resistance of 10 megohms on all ranges. Since this instrument is designed to process ac frequencies up to 1 MHz, each multiplier resistor

Figure 6–2. Simplified schematic diagram of TVM dc voltage circuitry.

is connected in parallel with a certain value of capacitance. These capacitances provide precise voltage division at both low and high frequencies. Observe that the bridge circuitry for the ac-voltage function is much the same as that for the dc-voltage function, except that a rectifier configuration is included between Q1 and Q2. It is instructive to analyze the operation of the bridge with this rectifier section in the circuit.

Observe in Fig. 6–3 that the incoming ac voltage passes first through Q1. This transistor functions as an electronic impedance transformer to match the high resistance of the multiplier to the low impedance of the meter circuit. Note that the output from Q1 is ac. This output branches into rectifiers D1 and D2 via C11. Full-wave rectification is thereby provided, or the rectifier output has a value proportional to the peak-to-peak voltage of the incoming signal. Note that diode D4 does not operate as a rectifier; it is used only for temperature compensation of the transistor characteristics. Next, the rectifier output is applied to R15, the ac calibration control. In turn, the rectifier voltage is applied to the gate of Q2, thereby unbalancing

the bridge circuit and producing a meter deflection that is proportional
to the gate voltage.

Next, a typical ohmmeter input circuit is shown in Fig. 6–4.
Note that the voltage drop across the 2000-ohm resistor will be a
function of the resistance under test. This voltage drop is applied to
the transistor bridge in the TVM, and the corresponding resistance
value is indicated on the ohms scale of the meter. Some TVM's also
provide a dc current function. As shown in Fig. 6–5, the milliampere
function does not employ the transistor bridge circuit. Instead, the
input current produces a voltage drop across shunt resistors, and this
voltage is applied directly to the meter movement. The shunt con-
figuration depicted in Fig. 6–5 is called a *ring shunt*. This arrangement
has the advantage that the shunt circuit remains closed while switch-
ing from one current range to another. In other words, there is no
danger of the full current from the test leads passing through the
meter movement while the switch position is changing. Of course, the

Figure 6–3. Simplified schematic diagram of TVM ac voltage function circuitry.

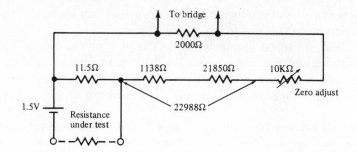

(a) Ohmmeter circuit R x 1 range

(b) Ohmmeter circuit R x 100 range

(c) Ohmmeter circuit R x 10000 range

Figure 6–4. Typical ohmmeter input circuit.

Figure 6–5. A milliammeter configuration with a ring shunt.

meter movement can still be damaged in a ring-shunt circuit, if the operator accidentally chooses a current range that greatly overloads the meter, and protective diodes are not provided.

6.3 TROUBLESHOOTING BASIC TVM SECTIONS

It follows from the foregoing discussion that the same circuit section in a TVM may operate on more than one function. As an illustration, the meter-movement section operates on the dc-voltage, ac-voltage, ohms, and milliampere functions. On the other hand, the ring-shunt section operates on the milliampere function only. Therefore, a block diagram for a TVM may be drawn in overlapping form, as exemplified in Fig. 6–6. Preliminary trouble analysis is facilitated by consideration of such functional diagrams. For example, if a TVM operates normally on all but the ohms functions, we would suspect that the ohmmeter battery is weak or dead. On the other hand, if a TVM operates normally on only the milliampere function, we would suspect that the supply battery is weak or dead.

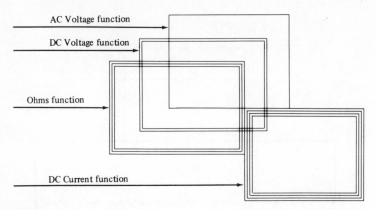

AC Voltage function

DC Voltage function

Ohms function

DC Current function

Figure 6–6. Block diagram of overlapping circuit sections in a TVM.

Compare the functional block diagram in Fig. 6–6 with the complete schematic diagram for a TVM shown in Fig. 6–7. Note, for example, that the same multiplier operates on the dc-voltage and ac-voltage functions. Thus, if the technician evaluates a particular group of trouble symptoms with respect to a functional block diagram derived from the complete schematic diagram, useful clues can often be deduced concerning the probable location of the fault. Common trouble symptoms include incorrect scale indication on one or more functions, one function "dead," more than one function "dead," inoperative ("dead") instrument, erratic indication, and intermittent operation. A TVM tends to drift slowly out of calibration as its components age. When the indicating error is slight or moderate, the TVM can be brought back into calibration by means of calibrating-control adjustments. On the other hand, if a substantial indicating error develops, it becomes necessary to locate the defective component, replace it, and then recalibrate the instrument.

With reference to Fig. 6–2, observe the 5 kilohm dc-calibrating control. This control is adjusted to obtain accurate indication of known dc-voltage values. There are two basic methods of obtaining known voltage values. First, if another TVM, VOM, or DVM (digital voltmeter) is available that is known to have ample accuracy, its indications can be cross-checked. Second, new mercury batteries are often used for reference. In other words, the voltage of a mercury battery remains close to its rated value of 1.35 volts until the battery is nearing the end of its life. For high-precision calibration checks, a standard Weston cell with a precision potentiometer arrangement may be employed. However, this latter method is both costly and time-

consuming; it is used principally for calibration of laboratory-type meters.

It is evident in Fig. 6–2 that adjustment of the dc calibration control will affect the scale indication on all ranges. If the multiplier is not defective, all dc-voltage ranges will be in calibration after any one range is brought into calibration. Accordingly, if the TVM is in calibration on one range, but is out of calibration on another or all other dc-voltage ranges, it is concluded that the trouble is being caused by one or more off-value multiplier resistors. Most service-type TVM's employ 1 percent multiplier resistors. Note that if R2 were 30 percent off-value, scale indication on the 1000-volt range would be in error accordingly, although indication on the 1-volt range would be practically unaffected. Observe also, that an apparent off-value multiplier resistor may actually be traced to a leaky capacitor. As noted previously, the same multiplier does double duty on ac and dc voltage functions, in this example. With reference to Fig. 6–3, it is evident that leakage in C2 will have the same effect on indication error as an incorrect value of R2.

After dc-voltage calibration has been completed, the ac-voltage calibration is checked. The most practical method is to cross-check the TVM scale indication on the ac-voltage function against the indication of another TVM, VOM, or DVM that is known to have ample accuracy. With reference to Fig. 6–3, R15 is adjusted as required to obtain correct indication on the ac scales. In case R15 is found to be out of range, or unresponsive, a systematic check should be made of C11, C16, D1, and D2. In other words, all of the dc-voltage function components are known to be normal at this point. In turn, the components utilized only on the ac-voltage function fall under suspicion.

As a practical troubleshooting note, Fig. 6–8 shows how a VOM can give misleading test results in open-circuit situations. Suppose, for example, that R10 or R13 in Fig. 6–3 has a cold-soldered joint or equivalent defect. Of course, the TVM is inoperative on all but its milliampere function. Next, if a 1000-ohms-per-volt VOM is used to check the voltage at the source terminal of Q1, and R10 is open, the TVM will resume normal operation and the source voltage will measure approximately the correct value when the VOM is operated on its 6-volt range. This result can be baffling to the beginner, who does not recognize that the 6000-ohm internal resistance of the VOM is substituting for the open 6.8 kilohms resistor. To repeat an important point, troubleshooting in solid-state circuitry is

Figure 6–7. Complete schematic diagram for a TVM. *(Courtesy of Sencore, Inc.)*

considerably facilitated by the use of instruments that have very high input impedance.

Next, with reference to Fig. 6–4, ohmmeter multiplier resistors are often damaged by accidental overload. In other words, the technician may forget that a circuit is "live" and attempt to make a resistance measurement with the TVM. In turn, one or more of the ohmmeter multiplier resistors is likely to be greatly changed in value from overheating, or burned out. Damage is usually confined to the

Figure 6–7. *(continued)*

ohmmeter input circuit, because the transistor bridge is protected by the neon lamp N32, and by resistor R13 in Fig. 6–2. Note that ring-shunt resistors as shown in Fig. 6–5 can also be damaged by overload. Unless the overload is particularly severe, the meter movement will be protected by diodes CR3 and CR4. These diodes normally do not conduct, but will go into forward conduction if the voltage drop across the meter movement becomes excessive. In turn, most of the incoming current is bypassed around the meter movement.

Figure 6–8. Internal resistance of the VOM shunts the open circuit.

6 . 4 BASIC OSCILLOSCOPE CIRCUITRY

Service-type oscilloscopes have operating controls as exemplified in Fig. 6–9. The functional sections are depicted in Fig. 6–10. A vertical amplifier is a high-gain wide-band arrangement with high input impedance. A typical vertical amplifier provides a voltage gain of 1500 at a bandwidth of 5 MHz, with an input resistance of 1 megohm. Such a vertical amplifier will produce a deflection of 1 inch on the CRT screen with an input signal amplitude of 0.02 volt. In other words, the CRT has a deflection sensitivity of approximately 30 volts per inch. Note that the time base generates a sawtooth output voltage which provides horizontal deflection of the electron beam in the CRT. A horizontal amplifier is utilized to step up the amplitude of the saw-tooth voltage sufficiently for full-screen deflection. Typical horizontal amplifiers have a voltage gain of about 50.

Note that the synchronizing section in Fig. 6–10 functions to keep the time-base oscillator locked in step with the vertical-amplifier signal waveform. In turn, the displayed pattern remains in fixed position on the CRT screen; if sync action is lost, the pattern will drift horizontally on the screen. A free-running time-base oscillator is depicted in Fig. 6–10. In other words, the time base operates continuously, whether a vertical-input signal is applied, or not. However, when a vertical-input signal is applied, the sync system feeds a sample of the signal waveform to the time-base oscillator. In turn, the oscillator frequency falls in step with the sync-signal frequency. The power supply has low- and high-voltage outputs. Low supply voltages are required by transistors, whereas a high supply voltage is required

Figure 6–9. Operating controls of a service-type oscilloscope.

for operation of the CRT. For example, a 5-inch CRT may operate from a 1500- or 2000-volt supply, with a current drain of less than 1 mA.

A basic cathode-ray tube circuit is shown in Fig. 6–11. A voltage-divider network is used to apply suitable operating potentials to each of the CRT electrodes. Control-grid bias is adjustable by means of the intensity control to set the level of pattern brightness. Operating potential on the focusing anode is adjustable by means of the focus control, enabling the operator to minimize the spot size on the screen. Direct current potentials on the vertical and horizontal deflection plates are also adjustable by means of the centering controls, so that the resting position of the spot can be changed as desired. In this arrangement, the deflection plates are at high potential, and the cathode of the CRT is near ground potential. On the other hand, many oscilloscopes operate the cathode at high negative potential, with the deflection plates near ground potential.

A typical vertical-amplifier configuration for a service-type oscilloscope is shown in Fig. 6–12. A signal voltage applied to the

Figure 6–10. Functional sections of a basic oscilloscope.

vertical-input connector is coupled through the frequency-compensated attenuator network. Note that capacitor C401 blocks dc flow when the vertical switch is in its ac positions. From the attenuator circuit, a portion of the input signal is coupled through resistor R1 and capacitor C1 to the gate of transistor Q1. Resistor R1 protects Q1 from being damaged in case a high potential is applied to the vertical-input connector while the vertical switch is in one of its lower positions. Diodes D1 and D2 are transistors connected to provide a Zener action. These diodes limit the input signal to approximately ±9 volts, which further protects Q1 from application of excessive gate voltage. Capacitor C1 improves the high-frequency response by providing a high-frequency path around R1.

Transistor Q1 is connected as a source follower. It provides a high input impedance and a comparatively low output impedance. Transistor Q2 is a constant-current source for Q1. Diodes D4 and D5 each provide a 0.6-volt drop (total of 1.2 volts) to hold the base of Q2 at a constant voltage. Since the circuit for Q2 is basically an emitter follower, and the emitter voltage is dependent upon the base voltage, the emitter voltage will also remain constant. This constant emitter voltage appears across resistor R2. Therefore, the cur-

Figure 6–11. A basic cathode-ray tube configuration.

rent through R2 is constant. Resistor R2 is adjusted so that the source voltage for Q1 is zero when there is no incoming signal.

A signal voltage applied to the gate of Q1 will cause only voltage changes at the source, because the current through Q1 is constant. These voltage variations are applied across the gain control R404, and a portion of this signal is applied to the gate of source follower Q3. Transistor Q4 forms a constant-current source for transistors Q5 and Q6. Since the emitter of each transistor is connected to this constant-current source, the current source serves as a common-emitter resistance and sets the operating point for the following stages. The output from source-follower transistor Q3 is amplified by Q5. A portion of the signal applied to the base of Q5 appears at its emitter. Because transistors Q5 and Q6 have a common-emitter resistance, the signal present at the Q5 emitter is effectively coupled to the emitter of Q6.

Transistor Q6 functions as a common-base amplifier, with its base voltage set by the vertical-position control R406. This control

Figure 6–12. Typical vertical-amplifier configuration. (*Adapted from a Heath Company original*)

positions the trace by applying a dc voltage to the base of Q6, causing a dc unbalance in the vertical amplifier. When the collector output voltage of Q5 decreases, its emitter voltage increases. An increased emitter voltage at Q6 reduces its forward bias and increases its collector output voltage. The signal at the collector of Q6 is 180 degrees out of phase with the signal at the collector of Q5, providing a push-pull amplifier action for driving the CRT deflection plates. Capacitor C3 is an emitter bypass capacitor that boosts the gain at high frequencies by means of partial-bypassing action. Emitter resistors R8 and R9 establish the dc gain of the vertical amplifier.

Driver transistors Q7 and Q8 operate as common-emitter amplifiers. In addition to providing gain, they also isolate transistors Q5 and Q6 from the output stages. Capacitor C4 provides ac coupling between the emitters of Q7 and Q8 and improves the high-frequency response of the stage. Output amplifiers Q9 and Q10 step up the push-pull signal to a suitable level for driving the vertical deflection plates of the CRT. In summary, the action of a vertical amplifier is much the same as that of an audio amplifier. The chief differences are that a vertical amplifier processes a very wide band of frequencies, so that frequency compensation must be employed. More stages are also required, because an amplifier that is designed for wide-band response necessarily develops comparatively low gain.

Next, it is instructive to consider the operation of the sweep-generator configuration shown in Fig. 6–13. Note that the sync switch selects either a portion of the vertical-input signal or a signal applied to the external-sync connector. In turn, this selected signal is then coupled to the gate of source-follower Q101. D102 and D103 are transistors operated as diodes to protect transistor Q101 from high voltages. In other words, D102 and D103 have Zener action at ±9 volts. Constant-current source Q102 is adjusted by the sync-level control R103 to provide proper biasing of the sync circuits. This ensures that even a small incoming signal will lock the sweep generator.

Transistors Q103 and Q104 amplify the sync signal and apply it to the Schmitt trigger circuit comprising Q105 and Q106. This Schmitt trigger circuit is a regenerative bistable circuit that generates a rectangular pulse each time that it is triggered and reset. Transistors Q109 and Q110 form an astable multivibrator arrangement. When Q110 is conducting and Q109 is cut off, one or more of the timing capacitors (C114 through C118) are charged through transistor Q110. As the voltage at the emitter of Q110 approaches the voltage at the base, as a result of the capacitor charge, Q110 will cut off and drive Q109 into conduction. The charged timing capacitor will now dis-

Figure 6–13. Sawtooth generator configuration for a service-type scope. (*Adapted from a Heath Company original*)

charge through the constant-current source circuit of Q111. Note that the setting of the frequency-vernier control R407 determines the current flowing through Q111, which, in turn, determines the discharge current (and discharge time) of the timing capacitor. As the timing capacitor discharges, a positive-going ramp voltage (sawtooth) is generated and is coupled to the horizontal amplifier. The frequency of the horizontal sweep is determined by the particular timing capacitor selected by the frequency-range switch and the discharge current.

Since transistors Q107 and Q109 have a common-emitter resistor, a signal applied to the base of Q107 is emitter-coupled to transistor Q109. The pulse output (sync signal) of the Schmitt trigger Q106 is coupled to Q109. This causes Q109 to turn on and Q110 to cut off, starting the sweep just prior to the time that it would normally begin. When the signal at the emitter of Q109 goes positive, a positive pulse is coupled through capacitor C107 to the base of blanking amplifier Q108. A negative-going output pulse is coupled through capacitor C109 to the grid of the CRT. This pulse turns off the electron beam during retrace, thereby preventing a visible retrace line from appearing on the CRT screen.

Next, Fig. 6–14 shows the configuration for a typical horizontal amplifier. Its operation is similar to that of the vertical amplifier depicted in Fig. 6–12, except that it lacks the Q7-Q8 stage. Note that the sensitivity of the horizontal amplifier is 0.1 volt per cm, whereas the sensitivity of the vertical amplifier is 0.03 volt per cm in this example. The bandwidth of the horizontal amplifier is 1 MHz, whereas the bandwidth of the vertical amplifier is 5 MHz. Less bandwidth is required in the horizontal amplifier, inasmuch as the sawtooth oscillator described above has a maximum repetition rate of 500 kHz. Note that the positive-going ramp voltage (sawtooth) from the sweep section is amplified and applied to the horizontal deflection plates of the CRT. Horizontal positioning is provided by R409, which varies the base bias voltage on Q206.

The associated power-supply and CRT circuitry in this example is shown in Fig. 6–15. Line voltage is connected through the slow-blow fuse and the on-off switch to the primary windings of the power transformer. A high-voltage secondary winding on the power transformer is connected to the voltage-doubler circuit comprising D301, D302, and C303. Capacitor C301 filters this negative high voltage, which is fed through resistor R416 to the grid of the CRT. The intensity and focusing voltages are also supplied to the CRT from the voltage-divider network consisting of R412, R413, R414, and R303. A separate 6.3-volt winding supplies the CRT filament voltage. Opti-

Figure 6–14. Typical horizontal-amplifier configuration. (Adapted from a *Heath Company original*)

Figure 6–15. Power-supply and CRT circuitry for a service type scope. *(Adapted from a Heath Company original)*

mum focus is obtained when the CRT deflection plates and the astigmatism grid are at the same potential. Since the vertical-deflection plate voltages (collectors of Q9 and Q10) are adjusted to 100 volts dc by the constant-current source Q4, the astigmatism voltage is also adjusted to 100 volts dc.

A low-voltage secondary winding is connected to the full-wave bridge-rectifier circuit comprising diodes D304, D305, D306, and D307 and capacitor C304. Zener diode ZD303 and resistor R312 maintain a constant voltage to the base of pass transistor Q301. The output voltage is regulated at 36 volts by the series pass transistor Q301 and Zener diode ZD303. By connecting equal loads from each side of the power supply to ground, two separate supplies are obtained at +9 and −9 volts dc. Another secondary winding is connected to the full-wave rectifier circuit consisting of diodes D312, D313, D314, and D315. Capacitor C305 filters the rectified voltage. Dropping resistors R315 and R314 provide output voltages of +150 and +180 volts dc, respectively. Resistor R313 and Zener diode ZD101 (on the sweep-generator circuit board) reduce the 180-volt source to a regulated +120 volts dc. A separate secondary winding supplies a one-volt peak-to-peak voltage to the front-panel output jack for calibrating the vertical sensitivity.

6.5 TROUBLESHOOTING OSCILLOSCOPE CIRCUITRY

Various trouble symptoms are encountered when malfunctions occur in oscilloscopes. As an illustration, the vertical amplifier may be unresponsive, or the displayed waveform may be clipped or otherwise distorted. Sometimes the pattern appears too far left or right on the screen, with the positioning controls out of range. Again, the focus occasionally becomes poor, or the beam intensity becomes weak. The sweep generator may develop malfunctions resulting in horizontal nonlinearity or loss of frequency control. Sync action may be lost. Some types of component defects cause the pattern to drift on the screen. Occasionally the vertical amplifier becomes unstable, with resulting oscillation and "blocking" of incoming signals. If the power supply fails, the entire scope becomes "dead." In case the intensity control is turned too high, the CRT screen will be burned, and permanent brown spots or lines will remain on the fluorescent screen.

A block diagram for the exemplified oscilloscope is shown in Fig. 6–16. When a trouble symptom occurs, the technician analyzes the control responses (Fig. 6–9) in order to obtain possible clues concerning the section that has become defective. For example, if the pattern is dim, and has poor focus with objectionable astigmatism, an attempt is first made to obtain a normal display by adjustment of the intensity, focus, and astigmatism controls. Then, if a normal display is not obtained, the technician measures the voltages at the CRT terminals. An incorrect value of high voltage indicates that the trouble is in the power supply. On the other hand, correct CRT terminal voltages indicate that the CRT should be replaced. The troubleshooting chart shown in Table 6–1 lists common trouble symptoms, and corresponding possible trouble areas.

Direct current voltage measurements are generally the most useful guide to pinpointing defective components. For example, normal dc voltages are shown in Fig. 6–12 at all terminals of the vertical-amplifier transistors. Because most of the circuits are dc-coupled, it is not entirely practical to compile a "cause and effect" troubleshooting chart for this type of scope. For example, a saturated transistor on one side of a differential amplifier may appear as a trouble condition on the other side. Note that turnoff tests of transistors can usually be made in dc-coupled stages. In other words, if the base and emitter of a transistor are temporarily short-circuited, the collector voltage normally jumps up to the supply voltage. (This type of test applies to bipolar transistors only.) It is often helpful to make signal-injection tests through the vertical amplifier with an audio-oscillator signal, using the CRT screen as an indicator. The same approach applies to the horizontal amplifier.

Remember that the high-voltage power supply in an oscilloscope can shock a person severely. Beginners sometimes forget that the heater, cathode, and control grid in the majority of scopes operate at a high potential above ground. When one is making voltage measurements in such circuits, it is important to use battery-operated instruments, because both test leads will be from 1000 to 2000 volts above ground (chassis or common-bus) potential. Either a battery-operated TVM or a VOM is suitable in this situation. When a vertical or horizontal amplifier is oscillating, the trouble will be found in bypassing or decoupling circuits. In other words, if a bypass or decoupling capacitor becomes open-circuited, sufficient positive feedback can occur to throw an amplifier into violent oscillation. In most cases, a systematic check of these capacitors must be made to locate

Figure 6–16. Block diagram of oscilloscope for troubleshooting reference.

TABLE 6-1

Troubleshooting Chart

Difficulty	Possible Area of Trouble
Neither pilot lamp nor CRT filaments light.	1. Fuse blown. 2. On-Off switch. 3. No ac power from outlet.
Pilot lamp lights, CRT filament does not light.	1. Power transformer. 2. CRT.
No spot or trace on CRT.	1. Positioning or intensity controls improperly adjusted. 2. High voltage power supply. 3. CRT.
Dot cannot be centered vertically.	1. Vertical Position control and associated circuit.
Dot cannot be centered horizontally.	1. Horizontal Position control and associated circuit.
No vertical deflection.	1. Vertical amplifier.
No horizontal deflection.	1. Horizontal amplifier.
Poor focus.	1. CRT. 2. Focus control. 3. Astigmatism control. 4. Resistors R412, R413, R414, and R303.
Trace acts erratic when the window is touched.	1. Clean the window with detergent to eliminate static charge.
Cannot synchronize input signal with sweep generator frequency.	1. Sync switch in the EXT position. 2. Control R103 misadjusted.
No retrace blanking or poor retrace blanking.	1. Transistor Q108. 2. Diode AZ101.
Pilot lamp changes intensity from bright to dim.	1. This is normal operation.

the defective unit. Note that if another oscilloscope is available, a check can be made for ac voltage across suspected bypass and decoupling capacitors. If the suspect is faulty, a high-amplitude ac voltage will be found across its terminals. On the other hand, if the capacitor is not defective, only a very small ac voltage will be dropped.

REVIEW QUESTIONS

1. Explain the basic operating features of a transistor voltmeter.
2. What is meant by the term calibration?
3. If a TVM is off-calibration over a part of its dc-voltage ranges, what component defect would you suspect?
4. When a TVM becomes off-calibration on its ohmmeter function, what would you check first?
5. In case a TVM becomes inoperative, except for its dc-current function, what would you check first?
6. If an ohmmeter is accidentally connected across a "live" circuit, what components are likely to be damaged?
7. Name the chief sections of a service-type oscilloscope.
8. State a typical voltage-gain figure and bandwidth value for a typical service-type oscilloscope.
9. How much accelerating voltage is commonly applied to the cathode-ray tube?
10. State a typical voltage-gain figure and bandwidth value for the horizontal amplifier in a typical service-type oscilloscope.
11. Describe a peak-to-peak calibrating procedure for the vertical amplifier in an oscilloscope.
12. What is a probable cause for waveform clipping in the vertical amplifier?
13. Would this same type of defect account for nonlinear horizontal deflection, if it occurred in the horizontal amplifier?
14. Why might the positioning controls drift out of normal range?
15. Give a reason for development of poor focus.
16. If the beam intensity becomes weak, what malfunction would you suspect?
17. In case the vertical amplifier breaks into violent oscillation and blocks signal passage, what component defect is most probable?
18. State a probable cause for loss of sync action.
19. How does the technician determine whether a CRT should be replaced?
20. Would a saturated transistor on one side of a differential amplifier sometimes appear as a trouble condition on the other side of the amplifier?

21. How are bypass and decoupling capacitors checked for suspected open-circuits?
22. Can transistor turn-off tests be made in dc-coupled circuitry?
23. Explain the appearance of an astigmatism trouble symptom.
24. What precautions should be taken when measuring the high-voltage power-supply potential in an oscilloscope?

Troubleshooting Solid-state Radio Receivers

7.1 GENERAL CONSIDERATIONS

Nearly all of the radio receivers in present-day use are of super-heterodyne design, as depicted in Fig. 7–1. In other words, the signal frequency from the antenna is heterodyned to an intermediate frequency. Most of the gain and selectivity is provided by the inter-mediate-frequency (IF) amplifier. Note that the mixer stage provides some gain at radio frequencies (RF). Following demodulation (detection) additional gain is provided at audio frequencies. These are the most basic sections of a superheterodyne radio receiver. Variations and elaborations will be found in practice, but the fundamental operating principle is the same in all receivers. One common design variation is seen in Fig. 7–2. Note that Q1 serves both as an oscillator and a mixer; this arrangement is called a *converter*. Observe also that

Figure 7–1. Basic superheterodyne radio receiver block diagram.

Figure 7-2. Schematic diagram for a small portable radio receiver.

153

CR1 functions as an overload diode to shunt excessively strong signals around the first IF stage.

A diode, CR2, is used as both a detector and an AGC (automatic gain control) source in Fig. 7–2. This diode is connected so that the setting of the volume control determines the dc operating point of the first IF stage, as well as the amount of audio signal power applied to the audio amplifier. A class-B output stage is utilized, with a tapped-speaker circuit. Note that the converter provides a gain of 30 dB, and IF amplifier 50 dB, and the audio amplifier 60 dB. The detector imposes a loss of approxiately 20 dB. Thus, the maximum over-all gain of the receiver is about 120 dB. Gain measurements may be made in radio troubleshooting procedures, as explained in greater detail subsequently.

7.2 PRELIMINARY TROUBLESHOOTING APPROACHES

Various kinds of trouble symptoms are encountered, such as "dead" receiver, weak reception, distorted output, interference (poor selectivity), oscillator dropout, erratic operation, and intermittent operation. When the technician makes preliminary checks of a "dead" receiver, localization tests are utilized. These tests include a visual inspection, measurement of the battery voltage, or power-supply voltage, measurement of the total current drain, speaker-circuit click test, local-oscillator test, and AGC test. A visual inspection will locate physical damage, such as broken fine wires from the loopstick antenna, breaks at the tuning-capacitor terminals, or breaks at circuit-board terminals. The loopstick should be inspected for cracks, and it should be verified that the tuning capacitor turns freely.

Leads to the volume control, speaker, earphone jack, and battery should be inspected for damage. Circuit boards should be carefully inspected for cracks. A board may be flexed at several points by applying light pressure with the eraser end of a pencil. If momentary return of sound output occurs, it is indicated that there is a crack in a printed-circuit conductor or a poor connection to a component lead. It is sometimes possible to pinpoint such defects without making electrical tests. It is good practice to check the battery polarity in battery-type receivers, because a replacement battery is sometimes accidentally reversed. In turn, it may be necessary to replace various transistors and electrolytic capacitors that have been damaged by the reversal of battery polarity. Note that if a battery weakens to

Figure 7–3. Measurement of total current drain by receiver.

the point that it supplies less than two-thirds of its rated voltage, the receiver is likely to become inoperative. Battery voltages must be measured under normal load.

Many component defects result in a change of the dc operating voltages in one or more stages. If the operating point of a single stage is greatly changed, the total current drain of the receiver is usually changed. This measurement is easily made, as shown in Fig. 7–3. Rated current drain is often specified in the receiver service data. Otherwise, the technician may be able to make a comparison test on a similar receiver that is in good working condition. An abnormal current drain generally indicates a short-circuited or leaky filter or decoupling capacitor. A subnormal current drain is likely to be caused by an open-circuited transistor or a break in the dc path of a stage. If the normal no-signal current drain is not known for a receiver, it can be estimated by the following rule of thumb. A converter stage will draw from 0.5 mA to 1 mA; each IF stage will draw about 1.5 mA, as will an audio driver stage; a class-B push-pull stage will draw approximately 4 mA. If a single class-A audio-output stage is employed, it will draw about 4 mA. Figure 7–4 summarizes these current drains.

If the current drain measures zero, there probably is an "open" in the supply circuit. On-off switches may become worn and defective, for example. If the no-signal current drain is as much as ±50 percent incorrect, there is probably a short-circuited decoupling capacitor or a very large change in the operating conditions of a stage. Direct current voltage measurements will then suffice in most cases to pinpoint the defective component. On the other hand, if the current drain is approximately normal, additional methods of testing are required. One of these methods is called the speaker-circuit click-test. It can

Figure 7–4. Current distribution for a typical 6-transistor receiver, under no-signal conditions.

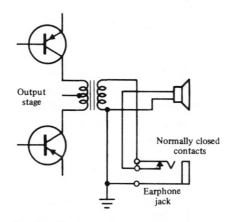

Figure 7–5. Speaker circuit is completed through the earphone jack.

localize a defect in the collector circuit of the output stage. The technician holds the speaker close to his ear, and listens for a click as the receiver is turned off and on. If no click is heard from the speaker, it is concluded that an open-circuit will be found in the speaker, output transformer, or associated connections. As depicted in Fig. 7–5, there may be a defective normally closed contact on the earphone jack, or a broken connection to the jack.

7.3 SYSTEMATIC TROUBLESHOOTING PROCEDURES

A "dead" receiver results from failure of the local oscillator. A useful quick check of oscillator operation can be made with another receiver that is in normal operating condition. Tune the other receiver to some

broadcast station in the range between 1MHz and 1.5MHz. Then place the two receivers near each other and turn the tuning dial of the "dead" receiver through the frequency range from 545 kHz to 1.45 MHz. If the oscillator in the "dead" receiver is operating, a loud whistle will be heard from the good receiver owing to interference from the oscillator. On the other hand, if no whistle (zero-beat action) is produced, it is indicated that the oscillator in the "dead" receiver is inoperative.

Next, consider the analysis of AGC trouble. Malfunction in the AGC sections can cause weak reception, overloading, or "dead receiver" trouble symptoms. In normal operation, the collector current of AGC-controlled IF stages varies from one milliampere to a few tenths of a milliampere, depending on the amplitude of the incoming signal. Note that this change in IF collector-current demand can be checked by monitoring the total current drain of the receiver. As an illustration, if a dip or dips are observed in total current drain while the tuning dial is rotated through its range, it is logical to conclude that the converter, IF amplifier, detector, and AGC sections are functioning (see Fig. 7–6). Therefore, the technician would turn his attention to the audio section in this situation. Of course, there must be a fairly strong local station signal available, or a generator signal must be utilized to make a definitive test.

On the other hand, suppose that peaks are observed instead of dips in current demand when the test is made as depicted in Fig. 7–6. It would be concluded that the peaks indicate the presence of audio drive to the class-B output stage. In turn, it is indicated that the complete signal channel is functioning, but that the speaker is not producing sound output. In other words, the trouble will probably be found in the speaker or in the secondary circuit of the output transformer (if used). As a practical note, remember that the receiver

Figure 7–6. Current-drain check for converter, IF amplifier, detector, and AGC section operation.

should be tuned slowly and carefully when making the foregoing current-drain test. If the tuning dial is turned rapidly, dips or peaks in current demand may be "skipped over." After the trouble has been localized to a general area of the receiver, additional tests must then be made to narrow down the possibilities.

7.4 LOCALIZING DEFECTIVE STAGES

At this point in the general procedure, the technician will usually have eliminated the local oscillator and the audio output circuitry from suspicion. In addition, the AGC test employing strong-signal and weak-signal inputs will generally have indicated whether the fault precedes or follows the detector section. Beyond this point, it is advisable to utilize signal-tracing or signal-substitution procedures to close in on the defective stage. Rough signal-injection tests can be made with a pencil-type noise generator, as illustrated in Fig. 7–7. The chief disadvantage of this method is lack of control over the injected signal level, and also the inability to tune the test signal to a specific frequency. On the other hand, if an audio oscillator or the audio output from a signal generator is utilized, the technician has control of the applied signal. To use an audio generator to inject signals into the audio section of a receiver, the ground side of the

Figure 7–7. Signal injection with a pencil-type noise generator.

generator is connected to the common bus (such as a battery terminal) in the receiver. A 0.1-μF capacitor is conected in series with the "hot" lead from the generator, to avoid drain-off of bias voltage in the audio circuits.

Signal-injection tests start at the collector terminal of the audio output stage, with the generator signal set to an amplitude that is just audible. Then, the signal-injection point is moved back through the audio section, as depicted in Fig. 7–8. Note that transformer winding ratios must be taken into account in signal-injection tests. As an illustration, when the signal-injection point is moved from (2) to (3) in Fig. 7–8, a definite decrease in sound output is noted from the speaker. This decrease is in consequence of the stepdown voltage ratio provided by T5. On the other hand, when the signal-injection point is moved from (3) to (4), a large increase in sound output normally occurs. Remember to maintain the generator output at a low but usable level as the signal-injection point is moved back stage by stage. In other words, the effects of transistor overload can mask the anticipated response.

Absence of response from a particular test point localizes the fault between the point where the response was last obtained and

Figure 7–8. Signal-injection points in a typical audio section.

the point where no output is produced. Thereupon, dc voltage, current, and/or resistance checks are usually made to pinpoint the defect to a particular component. In most trouble situations, only one component will be found faulty. However, if a leaky capacitor, for example, causes a transistor to draw excessive current, the transistor may also be found faulty. Note that current measurements can be made in the collector circuit of a transistor without taking any special precautions. In other words, the internal resistance of a service-type milliammeter will not upset the collector-circuit action. As an illustration, Fig. 7–9 shows a typical collector circuit that has a resistance of 2200 ohms effectively in series with an emitter resistance of 5600 ohms. In turn, the total resistance in this current path is 7800 ohms. If a 20,000 ohms-per-volt VOM is operated on its 10-mA range, the internal resistance of the instrument will be approximately 25 ohms. Since the addition of 25 ohms to a collector-circuit resistance of 7800 ohms is negligible, the current measurement will be accurate.

On the other hand, consider the problem of base-current measurement. With reference to Fig. 7–10, a typical transistor will draw 25 µA of base current in normal operation. In turn, a VOM

Figure 7–9. Current can be measured in the collector circuit without disturbing circuit action.

Figure 7–10. Base-current measurement will be in substantial error.

would be operated on its 50-μA range, and would have an internal resistance of about 2000 ohms. Making the reasonable assumption that the effective base input resistance is 1300 ohms, we recognize that insertion of the microammeter into the base circuit will more than double its total input resistance. Therefore, the base-current measurement will be in substantial error. For this reason, it is preferable to check base-circuit action with respect to voltage measurements. In various situations, collector-current measurements can be made in terms of voltage values. For example, in the circuit of Fig. 7–11, the technician customarily measures the drop across the 2200-ohm resistor and calculates the current flow from Ohm's law.

Next, if audio output is obtained with a low-level signal (about 3 mV) applied at the volume control, the audio section is eliminated from suspicion and it is indicated that the fault will be found in or ahead of the detector. Signal injection, using a modulated-RF voltage from a signal generator tuned to 455 kHz, is followed step by step as depicted in Fig. 7–12. A 0.01-μF capacitor should be

Figure 7–11. Ohm's law gives the collector-current value from the voltage drop across the 2200-ohm resistor.

connected in series with the "hot" lead from the generator to prevent shunting the dc bias voltage in circuits under test. Signal substitution through the IF section involves the same general procedure as described above for the audio section. If a weak signal (10 microvolts) applied to the base of the converter transistor produces an audible output, it is logical to conclude that the trouble will be found in the RF circuitry. Service-type signal generators seldom have calibrated output facilities. However, a 10-microvolt signal level can be established with the aid of a field-strength meter, as depicted in Fig. 7–13.

If the technician does not have an RF field-strength meter available, a comparative test can be employed to establish a suitable output level. In other words, a comparable receiver in good working condition is used for reference. A 455-kHz signal is applied to the base of the converter transistor in the good receiver, and the generator output is reduced until the audibility of the output decreases substantially. In turn, the generator output level is then correct for checking the receiver under test. To check the RF section of a receiver, it is good practice inductively to couple the test signal into the loopstick antenna. This method minimizes disturbance of the RF tuned circuits. A suitable coupling coil consists of a dozen turns of insulated wire on a 3-inch form. This coil is connected to the generator output cable and placed in the vicinity of the loopstick antenna. The generator is tuned to an RF frequency, such as 1000 kHz, and the receiver tuning dial is set to the same frequency.

Figure 7–12. (a) Progressive signal-injection test points *(Redrawn by permission of RCA)*

Figure 7–12. **(b)** Signal generator. *(Courtesy of RCA)*

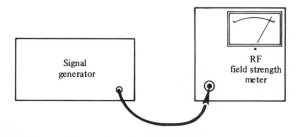

Figure 7–13. Field-strength meter checks generator output level.

7 . 5 SIGNAL-TRACING PROCEDURES

Technicians sometimes prefer to make signal-tracing tests instead of signal-substitution tests. Specialized radio signal tracers always provide speaker output, and may also provide meter indication. A typical signal-tracer configuration is shown in Fig. 7–14. Note that probes are provided, so that either high-frequency or low-frequency signals can be checked. Low-frequency signals are checked with the direct probe. It feeds the incoming signal into the audio amplifier of the instrument. Observe that a gain control is provided in order that amplifier overload can be avoided on high-level signals. When signal frequencies higher than the audio range are to be checked, a demodulator probe (RF probe) is employed. This type of probe functions as a traveling detector and responds to the wave envelope of the modulated-RF signal. Thereby, an audio signal is fed to the signal tracer.

Note that a supplementary built-in output transformer is included in the arrangement of Fig. 7–14. This transformer can be used with the internal speaker for tests of either single-ended or push-pull amplifiers. The tapped primary winding accommodates either transistor or tube amplifiers. In addition, the internal speaker can be used as a substitute speaker when one is troubleshooting audio equipment. Note also that a jack is provided in the signal tracer so that the internal amplifier can be used as a substitute amplifier in a radio receiver. Of course, a signal tracer must be used in combination with a suitable signal source. An AM generator is preferred, because it provides a steady signal with a controllable level. However, if the receiver under test is tuned to a broadcast station, signal-tracing tests can be made without difficulty.

Signal tracing proceeds from the antenna to the converter and step by step to the speaker. In other words, the detector probe is applied across the loopstick antenna terminals, and the volume control in the signal tracer is adjusted to obtain a sound output. The detector probe is then moved stage by stage toward the detector. Remember that coupling transformers usually step down the signal voltage from primary to secondary. To check the output from the detector and the signal levels through the audio amplifier, the audio probe is used with the signal tracer. Note that the meter in the signal tracer can be used for rough stage-gain measurements. However, these gain measurements are not entirely reliable in RF circuits, because the detector probe tends to load and detune resonant circuits.

Figure 7-14. Radio signal-tracer configuration.

7 . 6 TROUBLESHOOTING WEAK-OUTPUT CONDITIONS

A weak output trouble symptom is ordinarily analyzed by means of signal-injection or signal-tracing procedures. Preliminary analysis can be made by tuning to weak and strong stations. In other words, low volume on all stations indicates that the RF and IF sections are operating normally, and that the fault will probably be found in the RF section. On the other hand, if local stations are reproduced satisfactorily, but more distant stations cannot be tuned in, the defect will probably be found in the IF section. Some AGC malfunctions can cause the same trouble symptom. It is helpful to note typical stage gains in terms of signal voltage for a typical radio receiver. Thus, a converter stage normally provides a voltage gain of approximately 20 times; a diode detector imposes a loss of about 15 times; a transistor detector provides some voltage gain; the first IF amplifier has a gain of about 35 times, the second IF amplifier shows a gain of about 20 times; an audio driver provides a gain of approximately 35 times, and the audio output stage typically shows a voltage gain of 4 or 5 times. These signal-voltage relations are summarized in Fig. 7–15(a).

To measure stage gains in terms of signal voltage, the best procedure is to clamp the AGC bus at an average potential, so that the AGC section will not upset the measurements. An ac voltmeter is then connected across the speaker load, as depicted in Fig. 7–15(a). To measure the voltage gain of the audio-output stage, connect the output from an audio oscillator (with a blocking capacitor in the "hot" lead) across the voltmeter load. Then set the output from the audio oscillator to a low level that can be adequately indicated by the voltmeter. Note this meter reading. Next, move the signal-voltage point back to the output of the driver stage (input of the audio-output stage), and again observe the meter reading. If the amplifier gain is normal, the meter reading will increase from 4 to 5 times.

Intermediate frequency stage gain is measured in a similar manner. In other words, if an AM signal generator with a blocking capacitor connected in the "hot" lead is used, the gain of the second IF stage would be measured as follows. The ac voltmeter is left connected at the output of the audio section, and a 455-kHz modulated IF signal is applied at the output of the second IF stage. To avoid overloading, the signal level of the generator is reduced until a low but adequate indication is obtained on the meter scale. This reading is noted. Next, the signal is applied at the input of the second IF stage, and the meter reading is again observed. If the stage gain is

Figure 7–15. **(a)** Normal voltage gains for a small radio receiver; **(b)** A standard 5-transistor broadcast receiver configuration.

168

normal, the meter reading will increase about 20 times. Note that it might seem that a normal first IF stage has a low sensitivity, if the AGC bus is clamped to an incorrect potential. A correct clamp voltage will provide a base-emitter bias of 0.2 volt on the first IF transistor.

After a weak stage has been localized by means of gain measurements, the defective component can usually be pinpointed without undue difficulty on the basis of dc voltage measurements, supplemented by resistance measurements if necessary. Open coupling capacitors or open bypass capacitors are exceptions, and must be identified on the basis of signal-tracing tests. In other words, an open coupling capacitor will have signal present on the input end, but will be "dead" on the output end. On the other hand, a normal bypass or decoupling capacitor normally shows little or no signal voltage across its terminals. Presence of signal voltage indicates that the capacitor is open. As a practical note, the technician will find it very helpful to make comparison tests and measurements on a similar receiver that is in good working condition. Comparison tests often confirm or dispel suspicions that are quite difficult to resolve otherwise. A typical receiver is shown in Fig. 7–15(b).

7 . 7 DISTORTION ANALYSIS

Typical distortion trouble symptoms include mushy or raspy sound, whistles, motorboating (a putt-putt sound), squeals, and tweets. The majority of distortion symptoms occur in the audio section, although the trouble may be localized to the RF, IF, or detector sections. Preliminary evaluation of distortion symptoms can often be made on the basis of listening tests. In other words, the receiver operation should be checked on both high and low volume settings. If no change in distortion is noted, the fault is probably located ahead of the volume control. On the other hand, if the distortion increases as the volume control is advanced, the defect will probably be found in the audio section. If distortion appears to be the same at all settings of the volume control, check the sound output on weak stations and on strong stations. If the distortion increases on weak signals, the detector is likely to have a malfunction. As an illustration, the detector may have lost its forward bias. Normally, a small forward bias is applied to a diode detector to minimize the small-signal nonlinearity in demodulation. For example, note in Fig. 7–15 that the detector diode is forward-biased through approximately 300,000 ohms to the source voltage.

It is customary to employ a forward bias of approximately 0.1 volt across the detector diode. As a practical note, if a replacement diode is installed with reverse polarity, distortion will be evident on strong signals, owing to incorrect AGC action. As noted previously, if distortion increases as the volume control is advanced, the trouble will probably be found in the audio section. In many cases, the fault is an incorrect operating point for an audio transistor. For example, if a defective capacitor causes a shift in base-emitter bias, the audio waveform will be compressed or clipped, particularly at high volume levels. On the other hand, if the distortion becomes worse at low volume levels, the most likely cause is crossover distortion. This type of distortion has been noted previously on page 45, and is caused by lack of a small forward bias on class-B push-pull transistors.

Consider the situation in which tuning is very critical and the receiver breaks into a hiss or whistle as the tuning dial is turned. This condition indicates that regeneration is occuring in the RF or IF section. In such a case, the audio section is cleared from suspicion. Regeneration troubleshooting is discussed in greater detail subsequently. Now, we shall consider some details of audio-section troubleshooting for distortion symptoms. It is advisable to check collector voltage values at the outset; in other words, subnormal collector voltage can cause overload distortion, particularly on strong signals. Bias voltages should be checked at the driver and output transistors. Note that class-B output transistors should have a forward bias beteen 0.09 and 0.12 volt, to minimize crossover distortion. Whenever incorrect bias voltages are found, be sure to determine the cause and correct the fault. For example, an associated capacitor can be leaky, or a resistor may be considerably off-value. Sometimes trouble is caused by installation of an incorrect type of replacement transistor.

Next, it is instructive to consider the approaches to regeneration symptoms involving squeal, hiss, motorboating, and unusually critical tuning. Regeneration can result from unwanted positive feedback in a single stage, or around a loop that includes several stages. Preliminary analysis is often facilitated by rotating the volume control. In turn, if the squeal changes only in volume, but not in frequency, it is logical to conclude that the trouble is in the RF section, IF section, and/or AGC section. On the other hand, if the pitch of the squeal varies with the setting of the volume control, or if regeneration starts and stops at different settings, it is indicated that the audio section is included in the feedback loop. Very often, regeneration is the result of a poor (high resistance) ground connection. In turn, a

common impedance is produced, across which positive feedback may occur. Note in passing that some receivers will develop regeneration if the batteries become marginal.

Electrolytic decoupling capacitors are common offenders. As an illustration, capacitors such as C6, C13, and C14 in Fig. 7–15 are ready suspects in case of positive feedback difficulties. Similarly, C15 and C22 in Fig. 7–15 should be checked. Again, if regeneration is localized in the RF amplifier and AGC section, all ground connections should be inspected, and it should be verified that transformer shield cans are properly grounded. Bypass capacitors may become open, and are the next most likely cause of positive feedback. When oscillation occurs in a single IF stage, the neutralizing capacitor falls under suspicion. For example, observe the 4-pF capacitor in Fig. 7–16. If this capacitor becomes open, the stage will develop positive feed-back, and is likely to break into uncontrolled oscillation. Note that a replacement transistor with an excessively high beta value can also cause regeneration or oscillation.

In case one or more IF stages break into oscillation, a sub-stantial dc voltage will be measured at the detector output when there is no incoming signal. Since the effective beta of a transistor depends upon its bias, a receiver will sometimes squeal or oscillate on weak-signal reception, but will operate stably on strong-signal reception. In other words, the AGC voltage changes in accordance with the in-coming signal level. Sometimes an unstable IF stage can be brought under control by a small increase of neutralizing capacitance, as de-picted in Fig. 7–17. Approximately two inches of insulated wire are twisted to form a small capacitor. This "gimmick" is connected across the neutralizing capacitor, and a sufficient number of turns are formed

Figure 7–16. The 4-pf neutralizing capacitor prevents positive feedback.

Figure 7–17. A "gimmick" will increase the neutralizing capacitance.

to stabilize the stage. Avoid over-neutralization, as the stage gain will be reduced accordingly.

Distortion may be associated with excessive noise, or a receiver will sometimes provide distortionless reception with noise interference. Noise is a hiss or rushing sound; all receivers have a slight noise background, but it is normally unnoticed. When the noise level increases to the point that it impairs the entertainment quality of the audio output signal, the noise becomes a trouble symptom. Any stage, including the audio output stage, can produce large amounts of noise when a suitable malfunction occurs. Note that a noisy audio amplifier produces a hiss in the speaker output with a constant level that is not affected by adjustment of the volume control. On the other hand, if the noise level increases as the volume control is advanced, the trouble will be found prior to the volume control.

Excessive noise generation is usually the result of a defective transistor, particularly a transistor with excessive collector leakage. In turn, the defective transistor is easily pinpointed by dc voltage measurements. When collector leakage is excessive, abnormal collector current is drawn. It is ordinarily unnecessary to measure collector currents, since there is usually more or less series resistance in the collector circuit. Accordingly, abnormal collector current produces an excessive voltage drop across the series resistance, and the collector voltage in turn becomes subnormal. Of course, excessive noise is oc-

casionally caused by other component defects, such as a deteriorated volume control or fixed resistor. However, leaky transistors are the most probable sources of high noise levels.

7 . 8 ALIGNMENT PROCEDURES

Symptoms of misalignment include poor sensitivity, broad tuning (poor selectivity), and possible interference. As a general rule, alignment adjustments are made after all troubleshooting procedures have been completed. An exception to this rule is observed when it is known or suspected that alignment adjustments have been tampered with. Although receiver alignment tends to drift slightly after extended periods of time, substantial misalignment is caused by component defects. As an illustration, if C1 or C2 becomes open-circuited in the IF circuit shown in Fig. 7–16, the stage will become misaligned. Note that after a tuned-circuit component or its associated transistor is replaced, realignment is often required because of the change in shunt capacitance.

AM radio receivers are usually aligned by the peak-response method. In other words, alignment adjustments are made with respect to individual specified frequencies that are supplied by an AM generator. A weak generator signal is customarily utilized, so that the AGC section of the receiver remains essentially inoperative. That is, if a strong alignment signal were employed, the AGC section would automatically reduce the receiver output as the tuned circuits were brought into peak alignment, and the maximum response would tend to be masked. Unless the receiver service data specify otherwise, alignment procedure for an AM radio follows these consecutive steps:

1. Each tuned-IF circuit is peaked at the specified frequency (such as 455 kHz).
2. The oscillator trimmer capacitor (or slug) is adjusted for maximum receiver response at 1500 kHz on the tuning dial.
3. The oscillator padding capacitor (if provided) is adjusted for maximum response at 600 kHz on the tuning dial.
4. Each tuned RF circuit is adjusted for peak response at the frequency (or frequencies) specified in the receiver service manual.

A dummy antenna such as the one depicted in Fig. 7–18 is recommended by industry authorities for applying the test signal

Figure 7–18. Standard dummy antenna for AM broadcast receiver alignment.

Figure 7–19. Generator coupling-coil arrangement.

from a generator to the antenna-input terminals of an AM radio receiver. This dummy-antenna network provides a source impedance that approximates that of an average AM radio antenna. However, many technicians utilize a 200-pF series capacitor only as a dummy antenna. When the receiver employs a built-in loop antenna, the generator output can be inductively coupled to the loop, as shown in Fig. 7–19. If an elaborate receiver is to be aligned, it is advisable to follow the step-by-step procedure specified in the receiver service data.

REVIEW QUESTIONS

1. State the approximate stage gains for a conventional AM superheterodyne receiver.
2. Describe several faults that might turn up in a visual inspection of a malfunctioning receiver.
3. What is the usual result of accidental reversal of the supply-battery polarity?
4. How is a current-drain measurement made?
5. State the approximate current drains for each stage of an ordinary AM receiver.
6. If the current drain is zero, what fault would you suspect?
7. Give one possible cause for excessive current drain.
8. Explain how a "click" test is made in a radio receiver.

9. How can the troubleshooter quick-check the local oscillator in a receiver?

10. Describe a current-monitoring test for checking AGC action.

11. Discuss the various signal-injection tests that are ordinarily made.

12. Why is it impractical to measure the base current of a transistor with a VOM?

13. Can the output level from a signal generator be checked with a field-strength meter?

14. Explain how a generator signal is inductively coupled to a loop-stick antenna.

15. Briefly describe the functions of a signal tracer.

16. Outline the general signal-tracing procedure in a conventional radio receiver.

17. How is an AGC line clamped? How does the troubleshooter know when the value of clamp voltage is correctly set?

18. Describe the procedure that is employed when troubleshooting a weak-output symptom.

19. State the various types of distortion that are encountered.

20. How can preliminary distortion localization be made?

21. Discuss some common causes of audio distortion.

22. Explain the nature of regeneration trouble symptoms.

23. When regeneration occurs, what component defect would you suspect first?

24. Explain the four-step peak-alignment procedure for a radio receiver.

Troubleshooting Solid-state FM Mono Stereo Receivers

8.1 GENERAL CONSIDERATIONS

An FM monophonic receiver is similar in many respects to an AM radio receiver, as depicted in Fig. 8–1. Both AM and FM receivers have superheterodyne tuners. Intermediate frequency amplifiers are similar, although an AM IF amplifier operates at 455 kHz, whereas an FM IF amplifier operates at 10.7 MHz. Demodulation functions have the same end result in both types of receivers; however, the detector circuits are quite different in FM as shown in Fig. 8–2. In AM transmission the amplitude of the carrier is modulated, whereas in FM transmission the frequency of the carrier is modulated. Audio amplifiers are much the same in AM and FM receivers. Note, however, that few AM receivers employ high-fidelity audio amplifiers,

Figure 8–1. Comparative block diagrams for AM and FM receivers. **(a)** Basic sections in an AM receiver; **(b)** Basic sections in an FM receiver.

Figure 8–2. Two basic types of modulated waves. **(a)** Amplitude modulation; **(b)** Frequency modulation.

whereas most FM receivers provide high-fidelity reproduction. In turn, FM IF amplifiers have much greater bandwidth than conventional AM IF amplifiers. This increased bandwidth is required to process the comparatively high modulation frequencies utilized in high-fidelity reproduction. For example, a typical AM IF amplifier has a bandwidth of 10 kHz, whereas an AM IF amplifier has a bandwith of approximately 200 kHz, as depicted in Fig. 8–3.

Alignment is somewhat more critical in FM receivers than in AM receivers, because high-fidelity reproduction is generally desired,

Figure 8–3. Frequency response curve for FM IF amplifier.

in addition to good sensitivity and good selectivity. In other words, trouble symptoms in FM reception may involve either component defects or a drift in alignment adjustments. Note that drift is more of a problem in high-frequency tuned circuitry than in circuits that operate at comparatively low frequency. Thus, the RF circuits in an FM receiver operate in the range from 88 to 108 MHz, whereas the tuned circuits in a broadcast AM receiver operate in the range from 600 to 1500 kHz. As noted previously, FM IF circuits operate at 10.7 MHz, whereas AM IF tuned circuits operate at 455 kHz. Therefore, FM troubleshooting techniques are directed toward deciding whether a particular trouble symptom or group of trouble symptoms may result from a component defect or from resonant-circuit drift.

A service bench that is equipped to troubleshoot FM receivers should be provided with a TVM, oscilloscope, FM sweep and marker generator, plus the basic tools and equipment utilized in AM receiver servicing. A TVM that has a high-low ohmmeter function provides optimum testing capability. It is advantageous to use an oscilloscope that has at least 10-MHz bandwidth; in other words, a scope with adequate bandwidth and comparatively high sensitivity can speed up signal-tracing procedures. A scope also provides more information than a conventional signal tracer. An FM sweep generator should provide outputs both in the 100-MHz and in the 10-MHz range. As explained in greater detail subsequently, an FM sweep and marker generator may be combined with a stereo multiplex generator. A combination instrument is less costly than separate sweep and stereo generators.

8.2 PRELIMINARY ANALYSIS OF TROUBLE SYMPTOMS

Most of the more elaborate high-fidelity systems are AM-FM combinations. In turn, the troubleshooter has certain practical techniques available for preliminary analysis of malfunctions. As an illustration, if the receiver operates normally on its AM function, but is "dead" on its FM function, the technician turns his attention to the FM cir-

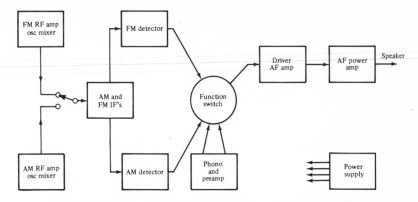

Figure 8–4. Block diagram for a simple AM-FM combo receiver.

cuitry. Thus, with reference to Fig. 8–4, the audio-amplifier section and the power supply can be eliminated from suspicion in this example. Note that the same IF-amplifier strip is utilized for both AM and FM reception. Therefore, the IF strip is dismissed from consideration in a preliminary analysis. In turn, the technician would turn his attention to the FM RF amplifier and oscillator-mixer circuitry. Note that switch contacts can become defective after long use, and this possible source of trouble should be kept in mind.

Following the same reasoning, if the power supply is not defective, but both the AM and FM sections are "dead," the technican looks for a defective component that is common to both AM and FM functions. As an illustration, an inoperative stage in the IF section of Fig. 8–4 could be responsible. Note also in Fig. 8–4 that a phono section is provided. In turn, this section can be switched into operation to determine whether the power supply and audio sections are functioning. In case all of the combination functions appear to be "dead," but the power supply is operating normally, it is good practice to check the volume control over its entire range. Sometimes a worn and "open" volume control will stop the audio signal unless it is turned to its maximum position. It is also advisable to press the function switch from side to side, and to push and pull on the control knob as it is rotated, inasmuch as an intermittent condition might become evident.

8 . 3 SIGNAL-INJECTION TESTS IN FM CIRCUITRY

A typical FM-AM receiver arrangement is diagrammed in Fig. 8–5. Note that the first and second IF stages process either FM or AM sig-

Figure 8–5. Signal-injection points and frequencies for a typical FM-AM receiver.

nals, whereas the third IF stage processes FM signals only. Let us assume that the receiver operates normally on its AM function, but is "dead" on its FM function. In turn, the technician will expect to find the trouble in either the FM detector, limiter, converter, or RF stages. Accordingly signal-injection tests are used to narrow down the possibilities in this example. To check the FM detector the procedure depicted in Fig. 8–6 is employed. A TVM is connected across C1 to measure the dc voltage that is developed. In some test situations, it is advantageous to disconnect one end of C1 and to connect the TVM across R1 and R2 to avoid limiting action. The first test is made with connections (1, 1), and the second test is made with connections (2, 2). An RF signal generator tuned to 10.7 MHz and with a blocking capicator in its "hot" lead is connected between the collector of the FM limiter and ground. If a reading is obtained on the TVM as the RF signal level is advanced, it is indicated that the ratio detector is functioning.

To determine whether the ratio-detector response is normal, the TVM is connected next at (2, 2) and is operated with zero-center indication. Then, as the frequency of the RF signal generator is varied through 10.7 MHz, the pointer should swing through zero from positive to negative values. In other words, the output voltage versus frequency should follow the standard S-curve depicted in Fig. 8–7. Next, consider the probable cause of no voltage in the (1, 1) test. No

Figure 8–6. Frequency modulation ratio-detector tests.

dc output indicates a malfunction in the secondary of the detector transformer, an open diode D1 or D2, or a "shorted" C1. Resistance measurements can be made for continuity checks, front-to-back ratio values, and for a short-circuited stabilizing capacitor. Note that transformer T1 could be seriously mistuned, but a dc voltage reading would nevertheless be obtained.

Next, consider the probable cause of failure to obtain both positive and negative readings in the (2, 2) test. This symptom throws suspicion on diode D1 or D2, leakage in C2 or C3, or misalignment of the secondary winding on T1. On the other hand, if the detector is operating properly, the next step is to check the IF strip. This signal-tracing procedure is the same as explained previously for AM receivers, except that a generator frequency of 10.7 MHz is employed. Note that the TVM is connected to the base of the limiter transistor if a discriminator is used for FM detector. However, if a ratio detector is used, the TVM is connected to the base of the third IF transistor. The general arrangement is depicted in Fig. 8–5. A 0.01-μF blocking capacitor is connected in series with the "hot" lead from the generator. A dc-voltage indication by the TVM shows that the signal is passing from the point of injection to the point of measurement. Some receivers may have as many as four IF stages, and each stage is checked in turn.

The converter stage is checked in the same manner as an IF stage. In other words, a 10.7 MHz generator signal is applied at the input of the converter transistor. If a dc-voltage indication is observed

10.7 MHz

Output - volts

−125 100 75 50 25 0 25 50 75 100 125+

Deviation-kilohertz

Figure 8–7. A ratio-detector S curve.

on the TVM, it is concluded that the converter stage is functioning. Next, the RF stage is checked with an RF signal in the range from 88 to 108 MHz, and with the receiver tuning dial set at the test frequency. Note that a dc-voltage indication on the TVM shows that both the RF amplifier and the oscillator stages are functioning. On the other hand, if no response is obtained, the trouble area must then be localized to either the RF section or to the oscillator circuit. Accordingly, the same test signal is applied at the input of the FM converter transistor. In case the TVM indicates an output voltage, the trouble will be found in the RF section. However, if there is no response by the TVM, the trouble will be found in the oscillator circuit. Direct current voltage and resistance measurements are usually made in the trouble area to pinpoint the defective component.

Some technicians prefer to use an oscilloscope to trace signals through an FM receiver. As noted previously, the scope must have adequate bandwidth if it is to display the signal directly. Since most shops utilize scopes with more or less limited bandwidth, a demodulator probe is often required. A demodulator probe, also called a signal-tracing probe, operates as a traveling detector. It processes an amplitude-modulated test signal as shown in Fig. 8–8. An RF AM signal is applied at the antenna-input terminals of the receiver, and the signal is traced as depicted in Fig. 8–9. If each stage is operat-

(a)

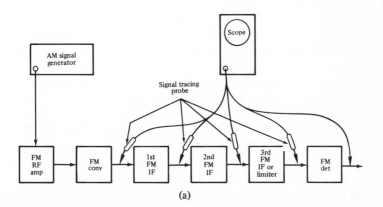

(b)

Figure 8–8. Demodulator probe for signal tracing. **(a)** Configuration; **(b)** Signal processing action.

(a)

(b) (c)

Figure 8–9. Signal tracing an FM receiver with an oscilloscope. **(a)** Procedure; **(b)** Input signal to IF stage; **(c)** Output signal from IF stage.

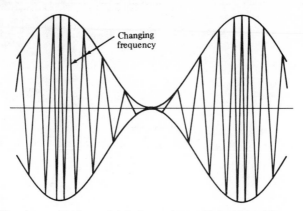

Figure 8–10. An amplitude-modulated waveform with incidental frequency modulation.

ing properly, an increase in amplitude of the scope pattern will be observed each time that the probe is moved forward a stage. Note, however, that a demodulator probe has a circuit-loading and detuning effect that makes it impractical to measure stage gains.

Observe in Fig. 8-9 that only an AM signal generator is utilized, although an FM detector arrangement is unresponsive to amplitude modulation. The reason that an output signal is developed by the FM detector results from the fact that service-type AM generators have appreciable incidental FM, particularly when operated at high percentage of modulation. In other words, service generators usually have amplitude-modulated oscillators, and this modulation arrangement produces FM as well as AM output (see Fig. 8–10). The most critical test point in Fig. 8–9 occurs at the output of the limiter stage. Complete limiting will result in a "dead stage" indication. Therefore, it is advisable to reduce the level of the test signal sufficiently that limiter action is more or less incomplete.

8.4 FM STEREO TROUBLESHOOTING

A stereo signal consists of two separate audio signals called left (L) and right (R) signals. In FM-stereo operation, the L and R signals are multiplexed (encoded) into the transmitted FM signal. Therefore, a multiplex decoder is required in a stereo receiver to separate the L and R audio signals from the composite stereo signal. In turn, the separated L and R signals are fed to L and R audio amplifiers. A block diagram for an FM-stereo receiver is shown in Fig. 8–11. In this ex-

ample, the multiplex unit comprises a bandpass amplifier, 38-kHz oscillator, detector, and matrix. Other multiplex decoding methods are also employed. However, the end result is the same in any case.

It is instructive to consider the waveforms that are involved in a stero-multiplex signal. With reference to Fig. 8–12, L and R speakers are utilized at the receiver, and these speakers correspond to L and R microphones at the FM transmitter. In this example, the outputs from the L and R microphones are being combined or added together to form an L + R audio signal. Note that this is essentially the same signal that would be produced by operation of a single microphone in an ordinary FM transmitter. In other words,

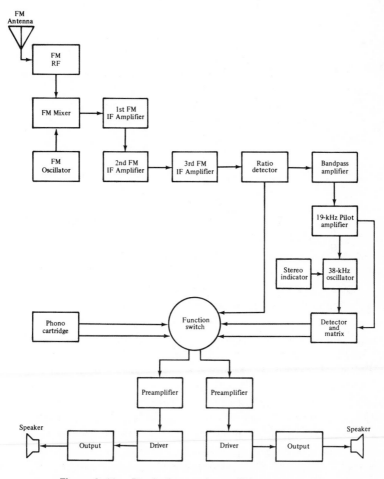

Figure 8–11. Block diagram for an FM-stereo receiver.

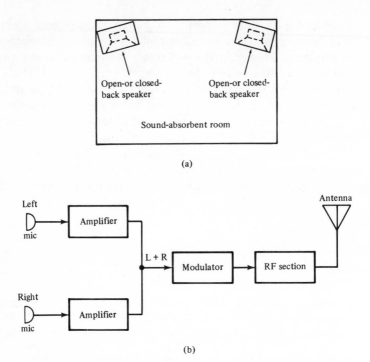

Figure 8–12. Microphones and speakers in a stereo system. **(a)** Typical speaker placement, at receiver location; **(b)** Microphones form an L + R signal at transmitter.

an L + R signal is equivalent to a monophonic audio signal. If this L + R signal is frequency-modulated on an FM carrier, it will be reproduced by a monophonic FM receiver in the same manner as if a single microphone were utilitzed at the transmitter.

Next, let us observe how signals are encoded in an FM-stereo transmission. As depicted in Fig. 8–13, a 38-kHz subcarrier is inserted with the mono or L + R signal. Since this subcarrier frequency is above the range of audibility, it does not affect transmission and reception of the L + R audio signal. Next, the 38-kHz oscillator is amplitude-modulated by another audio signal, called the L — R signal, as shown in Fig. 8–14. This L — R signal also occupies a frequency range from zero to 15 kHz, but since it is modulated on the subcarrier, it extends from 23 kHz to 53 kHz in the signal spectrum. In other words, the L — R signal is above the range of audibility and does not affect transmission and reception of the L + R signal.

It is instructive to observe the formation of the L + R and

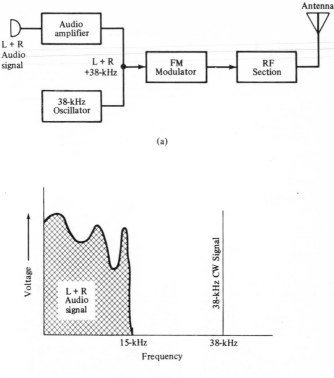

Figure 8–13. Insertion of the 38-kHz subcarrier. **(a)** Block diagram; **(b)** Frequency spectrum.

L — R signals from L and R microphones at the transmitter, because the decoder in the receiver simply operates "backward" in decoding the composite stereo signal. Figure 8–15 shows how the outputs from the L and R microphones are combined in a mixer to form the L + R signal. Note that the output from the R microphone is also passed through an inverter to form a —R signal. Then, this —R signal is mixed with the L signal to form the L — R signal. Thereby, the L and R signals are processed for encoding into the composite stereo signal. At the receiver, this waveform is decoded by specialized circuitry to recover the original L and R audio signals.

When the L + R and L — R signals are formed as shown in Fig. 8–15, the L — R signal is utilized to amplitude-modulate the subcarrier oscillator, and the resulting L—R modulated wave is mixed with the L + R signal, after which this composite stereo signal is

(a)

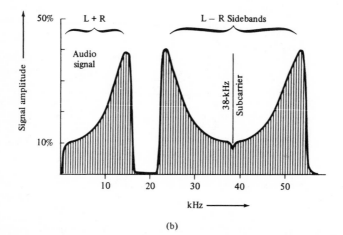

(b)

Figure 8–14. Modulation of the subcarrier by an L – R signal. **(a)** Block diagram; **(b)** Frequency spectrum.

frequency-modulated on an RF carrier for transmission, as depicted in Fig. 8–16. This composite stereo signal has the waveform illustrated in Fig. 8–16(b). Observe that this composite stereo signal has the R audio signal encoded as the upper modulation envelope, and has the L audio signal encoded as the lower modulation envelope. Accordingly, some FM-stereo receivers employ decoders that operate with envelope detectors to decode the composite stereo signal.

Note that the arrangement shown in Fig. 8–16(a) has been simplified for instructive purposes. In actual practice, a balanced modulator is employed to suppress the subcarrier and retain only the sidebands of the L — R signal. Subcarrier suppression is utilized in

Figure 8–15. Formation of the L + R and L − R signals from the L and R signals.

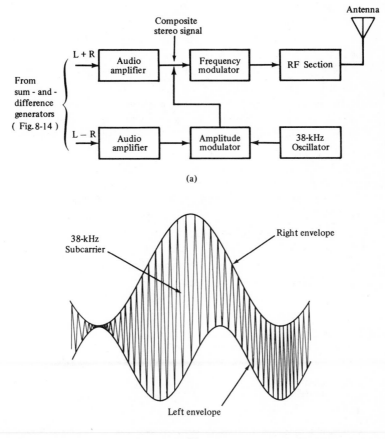

(a)

(b)

Figure 8–16. Modulation of the FM transmitter by the composite stereo signal. **(a)** Block diagram; **(b)** Composite stereo waveform.

order to reduce the energy content of the composite stereo signal, so that the optimum signal-to-noise ratio is realized. In turn, of course, the suppressed 38-kHz subcarrier must be reinserted at the stereo-FM receiver, in order to reconstitute the composite stereo signal. This is a subfunction of the multiplex unit (decoder) at the receiver. Since the 38-kHz subcarrier must be regenerated in precise frequency and phase by the multiplex unit, synchronization is required. This synchronization is provided by a low-level 19-kHz pilot subcarrier, as depicted in Fig. 8–17.

Three chief stero-multiplex decoding arrangements are used in various FM-stereo receivers, as depicted in Fig 8–18. In the bandpass-and-matrix arrangement, the L + R and L — R signals are added and subtracted in mixers to form the L and R signals. Note that the L — R sidebands are separated from the L + R signal by means of a 23- to 53-kHz bandpass circuit. Next, the output from the bandpass circuit is mixed with a 38-kHz frequency to reconstitute the L — R signal. An AM detector recovers the L — R audio envelope from the reconstituted L — R signal. Finally, the L — R signal is added to the L + R signal to obtain a 2L signal. Also, the L — R signal is subtracted via a phase inverter from the L + R signal to obtain a 2R signal. Note that the 38-kHz subcarrier must be synchronized by a 19-kHz pilot-subcarrier processing section (not shown in the diagram). A standard frequency-response curve for the bandpass unit is shown in Fig. 8–19.

In the electronic-switching method [Fig. 8–18(b)], the regenerated 38-kHz subcarrier alternately samples the positive and negative envelopes of the reconstituted composite stereo waveform. As seen in Fig. 8–16(b), these samples follow the contour of the L and R waveforms, whereby the composite signal is decoded. Finally, in the envelope-detection method depicted in Fig. 8–18(c), the recon-

Figure 8–17. Spectrum of the stereo-FM signal, as it arrives at the receiver.

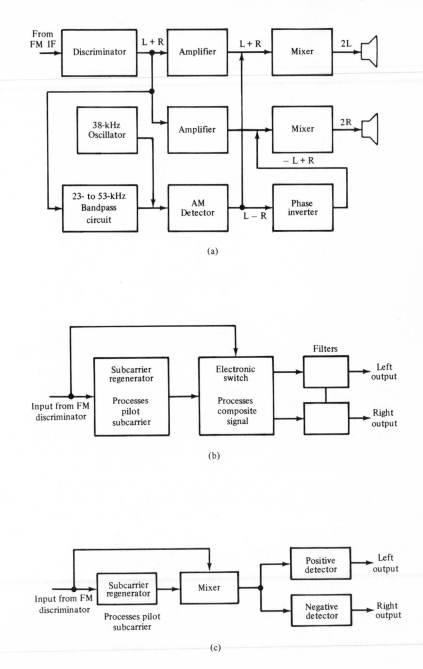

Figure 8–18. Basic types of stereo-multiplex decoders. **(a)** Bandpass-and-matrix arrangement; **(b)** Electronic switching method; **(c)** Envelope detection method.

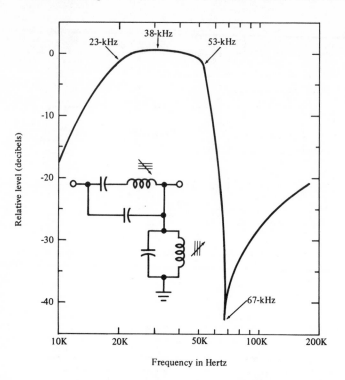

Figure 8–19. Standard frequency response curve for the bandpass filter circuit in a stereo decoder. *(Adapted from a Heath Company original)*

stituted composite stereo waveform is applied to a positive-rectifying detector and to a negative-rectifying detector. In turn, the composite stereo signal is decoded into its constituent L and R signals. In normal operation, any one of these decoding methods is capable of providing high-fidelity reproduction, with good separation of the L and R signals. As explained in greater detail below, various malfunctions can occur in this decoding process, and the technician must then make suitable tests to troubleshoot the decoding circuitry.

Figure 8–20 shows a configuration for a typical electronic-switching decoder unit. The composite audio input comprises the L + R signal mixed with the L — R sidebands and the 19-kHz pilot subcarrier. This input signal branches into a 19-kHz amplifier circuit and into the switching bridge comprising diodes CR3, CR4, CR5, and CR6. A frequency doubler (CR1 and CR2) follows the 19-kHz amplifier, in order to regenerate the 38-kHz subcarrier. In turn, the subcarrier is mixed with the incoming stereo signal to reconsitute the composite stereo waveform. Bridge conduction is determined by the

Figure 8–20. Configuration of an electronic-switching decoder unit.

instantaneous polarity of the subcarrier driving voltage, and the two sides of the bridge conduct alternately. Accordingly, the positive and negative envelopes of the composite stereo waveform are sampled alternately at a 38-kHz rate, and the bridge outputs consist of the decoded L and R signals.

The most basic test of a multiplex unit concerns the separation that it provides between the L and R signals usually expressed in decibels. Figure 8–21 shows the arrangement for a stereo-multiplex separation test. Signals L and R are applied in turn from a stereo-multiplex signal generator. A typical stereo signal generator with L and R meter indication for separation tests is illustrated in Fig. 8–22. Note that L and R signals have the waveshape depicted in Fig. 8–21(b), and differ only in the relative phase of their com-

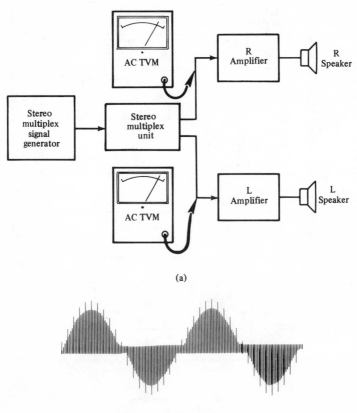

(a)

(b)

Figure 8–21. Stereo multiplex separation test. **(a)**Test setup; **(b)**Waveform supplied by stereo multiplex signal generator.

Figure 8–22. A stereo signal generator of the analyzer type. *(Courtesy of Sencore, Inc.)*

ponents. When an L signal is applied in a separation test, there is theoretically zero output from the R channel and maximum output from the L channel. Similarly, when an R signal is applied, there is theoretically zero output from the L channel and maximum output from the R channel. In practice, however, the separation is approximately 30 dB in normal operation. If the measured separation is 10 dB or less, a definite trouble symptom is recognized.

Inadequate separation is usually caused by a component defect in the multiplex unit, such as a defective capacitor, unmatched diodes, marginally operating diodes, leaky transistor, or off-value resistor. Note also that misalignment can cause poor separation; if a clean stereo waveform such as shown in Fig. 8–21(b) is not observed at the input of the decoder, check the FM receiver alignment. Note also that stereo frequency discrimination and poor separation can be caused

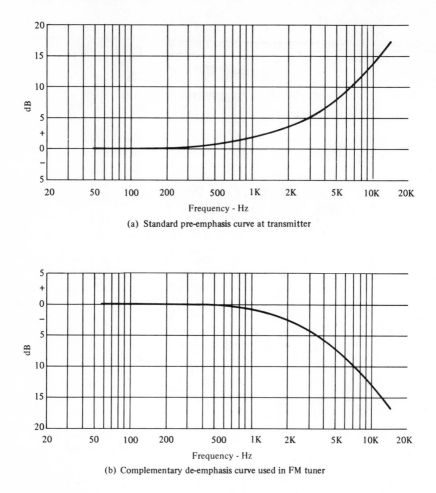

(a) Standard pre-emphasis curve at transmitter

(b) Complementary de-emphasis curve used in FM tuner

Figure 8–23. Frequency modulation preemphasis and deemphasis characteristics.

by incorrect connection of a decoder to an FM tuner. In other words, a separate FM tuner always has a deemphasis filter (RC integrating circuit) with a response as depicted in Fig. 8–23. Preemphasis must be used at the transmitter to obtain a good signal-to-noise ratio. In turn, deemphasis is required at the receiver. In a stereo system, it is essential to connect the deemphasis filter *after* the decoder.

Distorted output from a stereo decoder often results from poor synchronization of the regenerated subcarrier by the pilot subcarrier. There may be a defective capacitor in the switching bridge, a faulty electrolytic capacitor in the subcarrier regenerator section, or

Figure 8–24. Basic scope test points in a multiplex decoder unit. (*Adapted from an RCA original*)

197

a marginal transistor in an amplifier section. If the technician understands the waveforms that are normally present in a particular type of decoder, oscilloscope tests such as that depicted in Fig. 8–24 are very helpful in localizing the malfunctioning circuit. Stereo service data sometimes specify normal operating waveforms and peak-to-peak voltages. However, in many cases, the troubleshooter is thrown upon his own resources and expertise.

REVIEW QUESTIONS

1. State the two general sources of trouble symptoms and malfunctions in FM receivers.
2. Name the basic sections of a combination receiver.
3. How are sectional responses employed in preliminary trouble analysis?
4. Explain how a ratio detector is checked out.
5. If an oscilloscope has a vertical amplifier with 10 MHz bandwith, state the sections of an FM receiver that can be signal-traced using a low-capacitance probe.
6. Describe the function of a demodulator probe used with an oscilloscope that has comparatively narrow bandwidth.
7. What is meant by the term "incidental frequency modulation"?
8. Discuss the basic sections utilized in a stereophonic FM receiver.
9. Name the three chief stereo-multiplex decoding arrangements.
10. What is the most basic test of decoder operation?
11. How is separation checked, and in what units is it measured?
12. Is a separation of 10 dB normal in a stereophonic FM receiver?
13. How does an L+R signal differ from an L—R signal?
14. Define a pilot subcarrier.
15. Why is subcarrier suppression employed at the stereo-FM transmitter?
16. Why is preemphasis utilized at the stereo-FM transmitter?
17. If deemphasis were used prior to decoding at the stereo receiver, what malfunctions would result?
18. What is the audio-frequency range employed in stereo-FM broadcasting?
19. State the bandwidth utilized in an FM-IF amplifier.

20. What is the RF frequency range used in stereo-FM broadcasting?
21. Why is an L+R signal the equivalent of a monophonic FM signal?
22. Does the FM-IF amplifier in a stereo receiver have a 455-kHz center frequency?
23. Explain how an IF stage in a combination receiver can process either FM or AM signals.
24. Why is a blocking capacitor used in series with the "hot" lead of a generator in signal-tracing tests?

Troubleshooting Solid-state Quadriphonic Equipment

9.1 GENERAL CONSIDERATIONS

A quadriphonic sound system utilizes four speakers, as shown in Fig. 9–1. These speakers are often energized from four-channel quadriphonic tapes. When four separate sound channels are employed, the arrangement is called a discrete or true four-channel system. A discrete quadriphonic installation is an elaboration of a stereophonic system, in that the former comprises two additional speakers at the rear of the listening area. These rear speakers have the function of reproducing the reflected sound that is heard by an audience in a concert hall. Quadriphonic reproduction gives the reflected sound sources their approximately correct locations with respect to the listener. In addition to four-channel quadriphonic tapes, four-channel records are also available. These are essentially stereo recordings with encoded information in each of the stereo channels. A decoder is utilized with the quadriphonic amplifier system to "sort out" the four audio signals. Figure 9–2 shows standard track placements on stereo and quadriphonic tapes.

Although less favored than discrete quadriphonic sound, synthesized four-channel sound is in rather general use. A quadriphonic synthesizer operates in combination with two-channel records, tapes, or FM stereo to derive four-channel information from a two-channnel source. In other words, a two-channel stereo tape, for example, has some residual quadriphonic information mixed with the

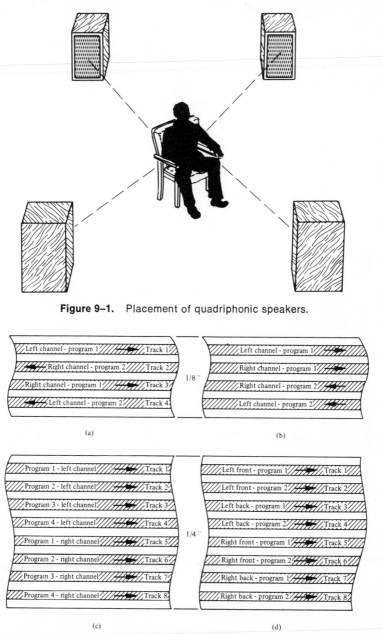

Figure 9–1. Placement of quadriphonic speakers.

Figure 9–2. Standard track placements on stereo and quadriphonic tape. **(a)** Reel-to-reel (open-reel) tape, stereo recording; **(b)** Cassette tape, stereo recording; **(c)** Eight-track cartridge tape, stereo recording; **(d)** Eight-track cartridge tape, quadriphonic recording.

salient stereo information. Moreover, stereo signals can be processed to simulate quadriphonic signals with some degree of realism. Thus, a synthesizer has the functions of recovery and simulation. Processing involves certain patterns of frequency discrimination, combined with specified phase shifts versus frequency. Linear processing circuitry is used exclusively in quadriphonic synthesizers. However, elaborate synthesizers include gain-riding circuits that improve sustained front or rear separation intervals. In other words, when a sensing network indicates that the rear-signal amplitude is somewhat greater than the front-signal amplitude, the gain-riding circuit automatically increases the amplitude ratio, thereby enhancing the front-rear separation. Conversely, a small dominance of front-signal amplitude is sensed and enhanced by the gain-riding circuit.

A four-channel adapter (synthesizer) is connected between a source of stereo signals and the four speakers in a quadriphonic system as shown in Fig. 9–3. Terminal-board connections are illustrated in Fig. 9–4. Note that FM broadcast stations cannot transmit discrete four-channel sound because of the required bandwidth for such transmissions. Regulations imposed by the FCC limit the bandwidth that can be employed. Figure 9–5 shows the specifications for commercial FM broadcast transmission. However, experimental quadriphonic FM transmission, using sophisticated encoding techniques, is being investigated. Possibly, these investigations will lead to developments that will place equivalent discrete quadriphonic broadcasting on a practical basis. In the meantime, development of more sophisticated synthesizers is also being actively pursued.

Disc recordings that provide four discrete channels of audio information have been produced, and are generally available. One

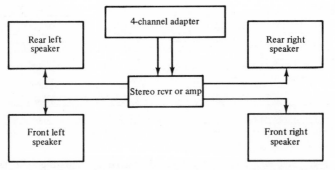

Figure 9–3. A four-channel synthesizer (adapter) is connected between a stereo-signal source and four speakers.

Figure 9–4. Terminal-board connections for a quadriphonic adapter.

technique utilizes a frequency-modulated supersonic carrier added to each channel of a stereo recording. The conventional L and R stereo channels provide the sum of the front and rear sound sources; they reproduce conventional stereo sound when played on conventional stereo record players. The supersonic carriers are modulated by the difference of the front and rear sound sources. In turn, after the carriers are demodulated and mixed with the R and L channel signals, four separate or discrete channels of audio are obtained. It is impractical for FM stations to transmit this particular quadriphonic signal, because it is excessively difficult to maintain adequate channel separation through the FM system.

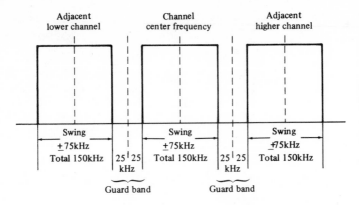

Figure 9–5. Specifications for FM transmission channel widths.

A block diagram of the system used in recording and playing back discrete four-channel tapes is shown in Fig. 9–6. A block diagram of the arrangement utilized in playing back four-channel discrete-quadriphonic records is seen in Fig. 9–7. The 30-kHz carrier is frequency-modulated by the stereo-difference signal during the recording process. Next, in the playback process, the audio signals from the two conventional stereo channels are mixed with the signals from the demodulated carriers. This mixing action reconstitutes the original four discrete audio signals, which are then amplified and fed to four speakers. Another approach to four-channel discrete reproduction employs encoders during recording and complementary decoders during playback, as shown in Fig. 9–8. Various types of encoder/decoder circuits are utilized. A widely used arrangement is depicted in Fig. 9–9. This method encodes audio signals by means of 180-deg phase shifts. It has the disadvantage, however, of cancellation of some audio frequencies in localized listening areas.

To avoid the foregoing 180 deg phase-shift disadvantage, the quadriphonic stereophonic (QS) system employs 90-deg phase shifts of encoded signals, as shown in Fig. 9–10. This QS system, in turn, has the disadvantage of incomplete separation of front and rear audio signals. To obtain improved separation, gain-riding circuitry is utilized in more elaborate versions of the QS system. These gain-riding circuits automatically exaggerate the separation between front and rear channels whenever a separation is sensed. Note that QS systems are compatible with conventional stereo equipment. However, the compatibility is not quite complete, and when a stereo record is played through a four-channel QS system, a synthetic diffused-stereo repro-

(a)

(b)

Figure 9–6. System used in recording and playing back discrete 4-channel tapes.

duction results. Some listeners approve of diffused-stereo reproduction, and others assert that the sound is unnatural.

Note that the final matrices are not shown in Figs. 9–9 and 9–10. In other words, the LF, LR, RR, and RF outputs are mixed in matrices to recover the left, right, front, and rear audio information for the encoded input signal. Thus, LF + LR = 2L; RR + RF = 2R; LR + RR = 2R (rear); LF + RF = 2F (front). Some quadriphonic installations employ four identical high-fidelity speakers with identical amplifiers. Again, other quadriphonic installations utilize high-fidelity equipment for L and R reproduction only, and use utility-type amplifiers and speakers for RR and LR reproduction. Utility-type equipment is less costly, and it may also be chosen on the basis that RR and LR sound makes little contribution to high-fidelity reproduction.

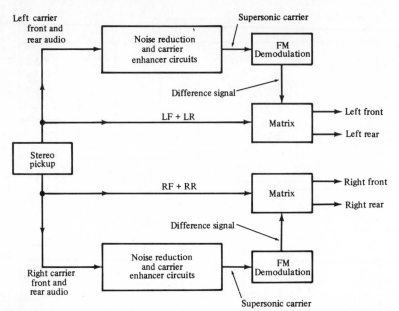

Figure 9–7. Arrangement used in playing back 4-channel discrete-quadriphonic records.

(a)

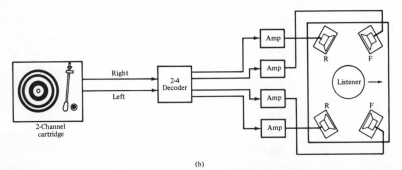

(b)

Figure 9–8. Encoder-decoder arrangement for 4-channel discrete reproduction.

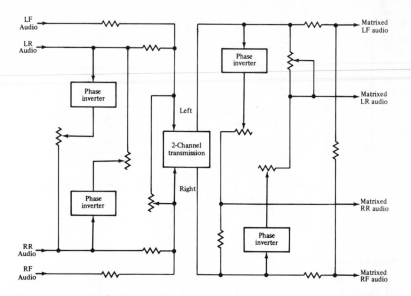

Figure 9–9. A 180° phase-shift encoder-decoder quadriphonic arrangement.

Figure 9–10. A 90° phase-shift encoder-decoder quadriphonic arrangement.

9.2 SQ QUADRIPHONIC DECODER CIRCUITRY

A schematic diagram for an SQ quadriphonic decoder utilizing three integrated circuits is shown in Fig. 9-11. The SQ system is a CBS development. Note that the first IC is the SQ matrix decoder. The

Notes:
Power supply +20V DC @ 75mA
C33, C34, C35, C36 values depend on
load of following stage. Use 0.3μF
for input impedance of 100000 ohms;
0.5μF for input impedance of 50000 ohms.
All resistors 1/4 watt 5% .
All electrolytics 25V or higher.

Figure 9–11. Schematic diagram for an SQ quadriphonic decoder.

Figure 9–11. *(continued)*

IC operates in combination with RC phase-shifting networks to provide decoding action. Next, the second IC is a four-channel voltage-controlled amplifier that accepts prevailing digital-logic commands and translates them into enhanced quadriphonic separation. Observe that the third IC develops the logic voltages that control the amplifiers in the second IC. The first IC comprises two preamplifiers that are fed with LT (left total) and RT (right total) signals, respectively. Each preamplifier, in turn, drives an all-pass network that generates two LT signals in quadrature and two RT signals in quadrature. Finally, matrixing action develops four separate audio signals that drive the speakers.

(a)

(b)

Figure 9–12. Layout of the SQ quadriphonic decoder diagrammed in Fig. 9–11. **(a)** Circuit board; **(b)** Parts placement.

Note that the second IC, which provides four voltage-controlled attenuators for varying the amplifier gains, can be set for relative gain between channels by means of three external dc voltages. This IC also operates with a master volume adjustment and with front/rear balance controls. It is evident that the third IC provides "gain riding" to improve the front/rear separation; it also provides variable dc logic enhancement (dimension control) for greater flexibility of the basic SQ decoder performance. This decoder is used only to play back quadriphonic records. The circuit board appears as shown in Fig. 9–12; a separate power supply is utilized which provides 75 mA at 20 volts. Note that the five maintenance controls are not mounted on the circuit board, but are connected via cable leads.

9.3 SPEAKER PLACEMENT AND ROOM ACOUSTICS

In addition to adjustment of the maintenance controls on a quadriphonic decoder, variations in the ambient sound effect may be obtained by different arrangements of the speakers, as depicted in Fig. 9–13. The optimum arrangement will depend upon room size, acoustics, and upon listener preference. Some rooms, such as rumpus rooms, tend to develop prominent reverberations. Other rooms, such as carpeted living rooms with heavy drapes, are comparatively "dead" from an acoustical standpoint. Therefore, a satisfactory speaker placement for a rumpus room is likely to be unsuitable for a living room. In many cases, it is preferable to place speakers in corners of rooms, inasmuch as a corner tends to direct the sound toward the center of the room. However, this is not an invariable rule, and improved ambience is sometimes obtained with speakers located halfway between one corner and the next.

Whenever the speaker arrangement is changed, the balance and dimension controls of the decoder should also be varied, to determine the optimum relation. It is helpful to have an assistant available, so that the changes can be promptly evaluated from a preferred listening position. Musical passages chosen for adjustment of a quadriphonic system should be representative of the listener's preference for entertainment. In other words, it is poor practice to adjust the system on the basis of demonstration quadriphonic recordings, or special-effects discs. Similarly, adjustments should be made on the basis of normal listening levels. Although there may be occasions on which an unusually high volume level would be appropriate, decoder adjust-

Figure 9–13. Variations in quadriphonic speaker arrangement. *(Adapted from an original by Howard W. Sams & Co. Inc.)*

ments and speaker placement should be made to conform with usual listening situations.

9 . 4 TROUBLESHOOTING QUADRIPHONIC DECODERS

As in the case of conventional stereo systems, separation is the most basic consideration in quadriphonic system operation. Separation tests of a quadriphonic decoder can be made to best advantage with a demonstration record that provides full sound output from each of the four speakers in turn. In case of poor separation, systematic tests are required to close in on the trouble area. IC sockets are utilized in the example of Fig. 9–12. In turn, a quick check for defective IC's can

be made by simply substituting a known good device for a suspected device. This testing procedure is analogous to "tube switching" in old-style electronic equipment. On the other hand, when a decoder has its IC's soldered into the circuit board, substitution tests are not feasible, and dc-voltage measurements are made at the outset. Note that if the decoder service manual does not specify normal dc-voltage values, it may be possible for the technician to obtain a similar decoder in normal working condition for comparison tests.

When normal dc-voltage values are unobtainable, the most logical preliminary approach is to check out the capacitors on the circuit board systematically, starting with the electrolytic capacitors. Since the internal circuitry of the IC's will usually be unknown, it is not possible to conclude whether an external capacitor may or may not be shunted by internal resistance of an IC. However, in case the IC's are plugged into sockets, it is easy to resolve this question by unplugging the IC's while checking the capacitors. This procedure will permit the in-circuit testing of various capacitors for leakage—for example, C21 or C22 in Fig. 9–11. On the other hand, capacitors such as C9 or C10 must be disconnected from their circuits before meaningful leakage tests can be made. As noted previously, either a foil conductor can be temporarily "razored," or one of the capacitor pigtails can be unsoldered from the circuit board.

In-circuit tests for open capacitors are made with the IC's in their sockets. A suspected open capacitor can be easily checked by "bridging" with a similar good capacitor, to determine whether normal operation is resumed by the decoder. It is essential to avoid accidental interchange of IC's on a decoder circuit board. Not only may IC's be carelessly inserted into wrong sockets, but an IC may be accidentally reversed end-for-end. Figure 9–12 depicts correct orientations of the IC's in this example. Incorrect insertion of IC's is very likely to produce a "tough dog" problem that could lead to excessive waste of time and effort. A similar pitfall for the beginner is replacement of electrolytic capacitors in wrong polarity.

A quadriphonic decoder such as that described above introduces less than 0.5 percent harmonic distortion in normal operation. However, it is not possible for the technician to make harmonic-distortion measurements on decoders, because quadriphonic signal generators are not generally available to the service trade. In turn, the technician is limited to listening tests; a listening test is meaningful if made on a comparison basis with other quadriphonic decoders. Objectionable distortion is usually caused by the same component and device defects that result in poor separation. Accordingly, the

same general troubleshooting procedure is followed as has been explained for poor-separation symptoms. In any case, it is important to avoid confusion between decoder distortion and record-player distortion. In case of doubt, a cross-check with another record player will provide a logical decision.

REVIEW QUESTIONS

1. Define the term "quadriphonic."
2. Distinguish between discrete and synthesized quadriphonic operation.
3. Name two sources of quadriphonic sound.
4. Why is a decoder required for reproduction of quadriphonic records?
5. Explain the general function of a quadriphonic synthesizer.
6. What is a gain-riding circuit?
7. Why do FM broadcast stations not transmit discrete quadriphonic sound?
8. State the established FM broadcast channel bandwidth.
9. What is the carrier frequency utilized in quadriphonic discrete-type records?
10. Name a disadvantage of the 180° phase-shift encoder.
11. Is there any disadavantage to the 90° phase-shift encoder?
12. Distinguish between the SQ and QS systems.
13. Are QS systems compatible with conventional stereo equipment?
14. Discuss synthetic diffused-stereo reproduction.
15. Does a quadriphonic system always employ four high-fidelity speakers?
16. Outline the operation of an SQ quadriphonic decoder.
17. Why must optimum speaker placement be experimentally determined in a quadriphonic installation?
18. Is it always desirable to place the speakers in the corners of a room?
19. Distinguish between "live" and "dead" rooms.

20. What is the most basic quadriphonic performance test?
21. Explain how separation is checked in a quadriphonic system.
22. Why is separation checked with suitable demonstration records?
23. Discuss the basis of comparison tests in decoder troubleshooting.
24. Is a quadriphonic decoder capable of operation at less than 0.5% harmonic distortion?

Troubleshooting Solid-state Industrial-Electronic Equipment

10.1 GENERAL CONSIDERATIONS

Solid-state industrial-electronic equipment utilizes nearly all types of devices, such as diode rectifiers, bipolar transistors, unipolar or field effect transistors, thyristors, thermistors, light-dependent resistors, Zener diodes, photoconductive cells, unijunction transistors, varicap diodes, varistors, light-emitting diodes, integrated circuits, operational amplifiers, and so on. These devices are used in power supplies, amplifiers, indicating instruments, control equipment, counters, sorters, inspection equipment, monitoring units, communication facilities, remote-control installations, process control systems, and related applications. Operational amplifiers (op amps) are a specialized type of integrated circuit. Op amps have made extensive inroads into industrial-electronic design, and are employed in numerous linear and nonlinear applications. Since op-amp technology differs considerably from conventional IC technology, it is instructive to consider the basic op-amp arrangements.

Op amps are used in precision instrumentation amplifiers, as signal conditioners, signal processors, signal generators, and in numerous special-purpose circuits. Signal-conditioning units include specialized integrators and differentiators, controlled current sources, logarithmic amplifiers, multipliers, dividers, and active filters. Signal-processing units include comparators, multiplexing and clamping amplifiers, absolute-value circuits, sample-hold and peak detector circuits, and rms converters. Signal-generating units include sine-

wave oscillators, square- and triangle-wave generators, ramp and pulse generators, and function generators. Special-purpose circuits include meter networks, frequency and phase measuring configurations, off-on controllers, proportional controllers, optimizing controllers, and semiconductor device test circuitry. Therefore, the following topics are chiefly concerned with understanding and troubleshooting op amps in typical industrial-electronics apparatus.

10 . 2 BASIC OP-AMP CHARACTERISTICS

Op amps are fabricated in various IC packages, such as illustrated in Fig. 10–1. Basically, a conventional op amp is a very-high-gain direct-coupled amplifier with two high-impedance inputs (typically 1 megohm), and a low-impedance output (typically 100 ohms). Practically all modern op amps are designed in IC form. A distinction is made between a conventional op amp (operational voltage amplifier, or OVA), and a related op amp called an operational transconductance amplifier (OTA). An OTA is essentially the same as an OVA, except that an OTA has a high output impedance (typically 5 megohms). Certain op-amp applications are facilitated by the use of OTA's instead of OVA's. Many op-amp designs are in use, and less elaborate designs are employed in less critical applications. In any case, effective troubleshooting of op-amp circuitry depends upon a good understanding of circuit action.

Figure 10–1. Appearance of typical operational amplifiers. *(Courtesy of Intech)*

The basic building block of an op amp is called the differential amplifier. A differential amplifier consists of a pair of transistors, as depicted in Fig. 10–2. They are connected so that the configuration responds to the difference between two input voltage or currents. During this response, the differential amplifier suppresses input voltages that have the same amplitude and the same polarity. From a general viewpoint, a differential amplifier can be regarded as a push-pull amplifier. With reference to Fig. 10–2, Q1 and Q2 comprise the differential-amplifier arrangement, which is current-fed from the controlled constant-current source Q3, called a transistor current sink. Note that Q3 functions as a constant-current source because it operates at fixed bias and its collector-current flow remains practically constant as its collector voltage swings up or down.

Note that there is a precise balance between input terminals 1 and 5 in Fig. 10–2, because the base-to-emitter voltages and current gains (beta) of Q1 and Q2 are closely matched. An op amp contains cascaded differential amplifiers, as exemplified in Fig. 10–3. The output stage consists of Q6 and Q7. Regardless of its internal circuitry, an op amp is always represented by a triangular symbol, as depicted in Fig. 10–4. The two input terminals are called the inverting (−) input, and the noninverting (+) input. Most op amps have a single-ended output. Both the output terminal and the two input terminals are referenced to ground; this ground connection is the common point of the return circuit for the supply-voltage sources. In other words, two batteries may be employed as voltage sources; one battery applies a positive potential to the (V+) terminal of the op amp, and the other

Figure 10–2. Basic IC differential-amplifier circuit.

Figure 10–3. An op amp contains cascaded differential-amplifier stages.

battery applies a negative potential to the (V—) terminal of the op amp.

Most op-amp schematics do not show the voltage sources explicitly, and indicate an op amp simply as a triangle with two input terminals and one output terminal, as depicted in Fig. 10–4(b). As a practical note, an op amp can be ruined if the voltage-source polarity is accidentally reversed. Therefore, if this possibility exists, a pair of protective diodes may be included in the source leads, as shown in Fig. 10–5. These diodes permit forward-current flow, but will block reverse-current flow. Next, the input and output terminals of an op amp may be operated without negative feedback. However, in the vast majority of applications substantial negative feedback is provided. In turn, it is instructive to observe the basic characteristics of the circuit shown in Fig. 10–6. Part of the output signal is fed back to the input in opposite polarity. In turn, part of the input signal is cancelled out. Note that the gain of the amplifier is not reduced—on the other hand, the gain of the system is reduced.

Next, it is helpful to note that since the gain of an op amp is extremely high, such as 1000 times, and the system gain with nega-

Figure 10–4. Schematic symbols for an op amp. **(a)** Symbol showing power-supply connections; **(b)** Conventional symbol without power-supply indication.

tive feedback is usually very much less, it can be considered in the first analysis that 100 percent negative feedback is being employed. Practically, this means that the input resistance of the arrangement depicted in Fig. 10–6 is the same as the resistance of the input component. This conclusion follows from the fact that if 100 percent negative feedback were used, the inverting-input terminal of the op amp would remain at zero potential, regardless of the amplitude of the input signal. In other words, we may regard the inverting-input terminal as a virtual ground, in this example. Therefore, the input resistance to the op-amp arrangement becomes equal to the resistance of the input component. Specific examples of input components will be described in the next section.

Another practical conclusion concerning the majority of op-amp applications employing negative feedback is that the system gain can be regarded as equal to the ratio of feedback resistance to input resistance, as exemplified in Fig. 10–7. Next, observe that an op amp can be driven either at its inverting input terminal, or at its

Figure 10–5. Diodes D1 and D2 prevent damage to the op amp if the voltage-source polarity is reversed.

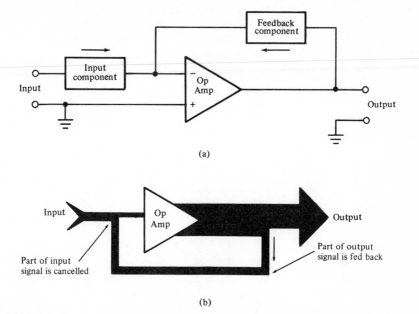

(a)

(b)

Figure 10–6. Basic op-amp feedback arrangement. **(a)** Block diagram; **(b)** Signal relations.

noninverting input terminal, as shown in Fig. 10–8. If it is driven at its inverting-input terminal, the output waveform will be shifted 180 deg in phase. On the other hand, if it is driven at its noninverting-input terminal, the output waveform will be in phase with the input waveform. Observe in Fig. 10–8(b) that the input impedance is extremely high because the input signal is not being applied directly against the feedback voltage. However, the effective negative feedback is practically the same in either the inverting or the noninverting configuration.

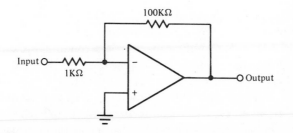

Figure 10–7. An op-amp configuration with an input resistance of 1000 ohms and a gain of 100 times (approximately).

Figure 10–8. Basic single-ended input circuitry. **(a)** Inverting-amplifier configuration; **(b)** Noninverting-amplifier configuration.

10 . 3 OP-AMP DIFFERENTIATORS AND INTEGRATORS

An outstanding advantage of op-amp circuitry is its ability to perform precise mathematical operations on electrical waveforms. Consider the simple RC differentiating circuit shown in Fig. 10–9. When a square-wave voltage is applied to the input of the circuit, a pulse waveform appears at the output. These pulses are a rough approximation to a true mathematical derivative of the square wave. Next, observe the relations depicted in Fig. 10–10. This series RC circuit provides both differentiating and integrating action. Note that the voltage waveform across the capacitor is a rough approximation to a true mathematical integral. With reference to Fig. 10–11, a true mathematical derivative of a square wave consists of a train of impulses, and a true mathematical integral of a square wave consists of a series of triangular waves (back-to-back sawtooth waves).

Note that a true impulse waveform has infinitesimal width and infinite amplitude. Of course, an ideal impulse cannot be realized by

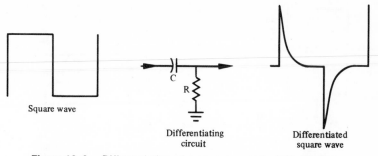

Square wave

Differentiating
circuit

Differentiated
square wave

Figure 10–9. Differentiation of a square wave by RC circuitry.

circuit action. However, a far more precise approximation to an impulse waveform can be obtained with an op-amp differentiating circuit than with a simple RC differentiating circuit operating on a square-wave input voltage. Consider the action of the circuit depicted in Fig. 10–12 when an input square waveform is applied. This configuration comprises a conventional RC differentiating circuit, with the input and output terminals of an op amp connected across the resistor. In turn, the capacitor feeds into a virtual ground point, instead of a resistive load. This means that the time constant of the input circuit is practically zero, so that the inverting input of the op amp is driven by extremely narrow pulses. In turn, these approximate impulses appear in amplified and inverted form at the output terminal of the op amp.

Next, consider the action of the op-amp integrating circuit shown in Fig. 10–13. The configuration comprises a conventional RC integrating circuit, with the input and output terminals of an op amp

(a)

(b)

Figure 10–10. Differentiated and integrated waveforms. **(a)** RC circuit driven by square-wave generator; **(b)** Universal time-constant chart.

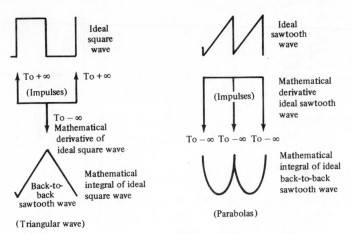

Figure 10–11. Ideal mathematical derivatives and integrals.

Figure 10–12. Op-amp differentiating circuit.

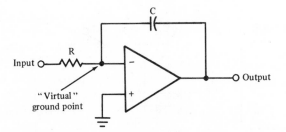

Figure 10–13. Op-amp integrating circuit.

connected across the capacitor. The circuit action can be understood easily by recognizing that the feedback capacitor places the inverting input of the op amp at virtual ground potential, at least in the first analysis. Therefore, if a dc input voltage is applied, a constant input current will flow while the feedback capacitor is charging up to the

applied potential. When a capacitor is charged by a constant current, the terminal voltage of the capacitor rises linearly. Thus, if a square-wave voltage is applied to an op-amp integrator, a triangular output waveform is obtained, as depicted in Fig. 10–11. Note that the basic differentiating and integrating circuits shown in Figs. 10–12 and 10–13 are not entirely practical, owing to the possibility of op-amp malfunction or damage. Means of avoiding these trouble points are explained subsequently.

10 . 4 OP-AMP SUMMING AND SUBTRACTING CONFIGURATIONS

Many industrial-electronic op-amp applications employ summers and subtracters, such as exemplified in Fig. 10–14. A basic summer comprises an op amp with considerable negative feedback, and a multiplicity of identical inputs to the noninverting terminal as depicted in (a). Since each input resistor feeds into the virtual ground point, these inputs are isolated from one another. Each input has an input resistance of 40 kΩ in this example. The gain of the summer from an input terminal to the output terminal is unity. In other words, no voltage gain is provided, although considerable power gain is developed owing to the low output impedance of the op amp. Observe that this arrangement will operate as a subtracter, if one of the input voltages is reversed in polarity. Note also that ac voltages can be added by a summer, provided that the inputs have the same frequency and phase. The input voltages must also have the same waveform.

A basic op-amp subtracter arrangement is shown in Fig. 10–14(b). Both the inverting and noninverting inputs are utilized. In this example, a voltage gain of 100 times is provided by the system. Note that this arrangement will operate as an adder, if one of the input voltages is reversed in polarity. Alternating current voltages can be processed, provided that the input voltages have the same frequency, waveform, and phase relation. Industrial electronic-control units often utilize a more general form of op-amp summer, called the scaling summer. A basic configuration for a scaling summer is depicted in Fig. 10–14(c). It provides scaling (weighting) of each input before it is added to the other inputs. As an illustration, R1 may have a value of 1 kΩ, R2 a value of 2 kΩ, and R3 a value of 3 kΩ. Then, if E1, E2, and E3 have equal values, R1 conducts twice as much current as R2, and three times as much current as R3. In turn, a relative output of 6E3 is obtained.

Virtual
ground

Output =
(Input 1 + Input 2
+ Input 3)

(a)

Output =
(Input 1 − Input 2)

(b)

Output =

$$R_o(\frac{E_1}{R1} + \frac{E_2}{R2} + \frac{E_3}{R3})$$

(c)

Figure 10–14. Op-amp summing and subtracting arrangements. **(a)** Basic summer; **(b)** Basic subtracter; **(c)** Basic scaling summer.

10 . 5 OP-AMP MULTIPLYING AND DIVIDING CONFIGURATIONS

Various industrial-electronic control circuits require multiplying or dividing operations to be performed on voltage or waveform data. Multiplication is accomplished by converting input signals into their logarithmic values by means of logarithmic amplifiers. These logarithmic values are then added together, after which the product is formed by action of an antilog amplifier. Figure 10–15 shows the basic multiplier arrangement. A logarithmic amplifier has an output that is proportional to the logarithm of the input. Note that the voltage across feedback transistor Q1 in (a) depends logarithmically on the feedback current. In other words, the emitter-base characteristic of a bipolar transistor is logarithmic over a portion of its operating in-

Figure 10–15. Basic op-amp/transistor multiplier configuration. **(a)** Skeleton schematic of logarithmic amplifier; **(b)** Skeleton schematic of op-amp/transistor antilog amplifier; **(c)** Block diagram of analog multiplier arrangement.

terval. Since the feedback current is proportional to the input current, the output of the logarithmic amplifier is proportional to the logarithm of the input voltage. Note that a power transistor is used as a feedback component so that the series feedback resistance is comparatively low.

Next, a basic antilog amplifier is depicted in Fig. 10–15(b). A grounded-base bipolar transistor is utilized as the input component, and a resistor is used as a feedback component. If Q1 is operated over a suitable interval of its emitter-base characteristic, the output from the op amp will be proportional to the antilog of the input current. In turn, if a pair of log amplifiers are employed to feed a summing amplifier, which is followed by an antilog amplifier, as diagrammed in Fig. 10–15(c), the input voltages A and B will be multiplied, and an output proportional to their product will be provided by the antilog amplifier. It will be recognized that if the summing amplifier is replaced by a subtracting amplifier, the output from the antilog amplifier will be proportional to the quotient of the inputs A and B. In other words, the difference between log A and log B is equal to the log of A/B, and the antilog of this quantity is equal to A/B.

Practical maintenance and troubleshooting procedures involve correct adjustment of the 10 kΩ potentiometer indicated in Fig. 10–15(a). This is called an *output offset null adjustment*. The more elaborate types of op amps are designed to have practically zero volts appearing at the input terminals and at the output terminal when no signal is present. This is an important consideration, because when a resistive feedback loop is employed, any residual output voltage (offset voltage) will be fed back to the input, and will cause nonlinear operation or other malfunction of the op amp. Many op amps have internal nulling provisions, and the 10 kΩ potentiometer in Fig. 10–15(a) provides an adjustable output offset null circuit. This "nulling" potentiometer is precisely adjusted by the technician to obtain zero volts output from the op amp when no input signal is being applied.

10 . 6 DIFFERENTIAL OUTPUT OP AMPS

Only single-ended output op amps have been discussed above. However, double-ended output (differential-output) op amps are also employed in industrial-electronic servo motors, and similar applications. Figure 10–16 shows the basic configuration for a differential-output type of op amp. Note that two negative-feedback loops are utilized. A differential-output op amp may be used to convert a

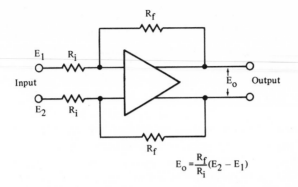

Figure 10–16. Basic configuration for a differential-output type of op amp.

single-ended input voltage into a pair of balanced output voltages (push-pull voltages) by grounding either one of the signal-input terminals. It is evident that a differential-output op amp can be a summer or subtracter. An important feature of this type of op amp in some applications is the fact that both of the input terminals and both of the output terminals are above ground potential, or "floating." The same considerations of output offset nulling apply to differential-output op amps as to single-ended output op amps. Note that in comparatively uncritical circuitry, output offset null adjustment may not be required.

10 . 7 TROUBLESHOOTING OP-AMP INSTALLATIONS

When trouble symptoms develop in op-amp installations, the first question to be resolved concerns, "Is it the op amp(s), or is it an external component that is causing the malfunction?" The answer can be obtained quickly, provided that the op amp(s) can be unplugged and replaced. On the other hand, if the device terminals are soldered into a circuit board, preliminary analysis and test procedures are required. The technician should not confuse op-amp system trouble with power-supply malfunction. Drift or ripple in power-supply voltage can cause puzzling trouble symptoms in op-amp operation. Therefore, it is advisable to check the power supply with an accurate dc voltmeter, such as a DVM, at the outset. If the power supply is adequately regulated and stable, check the output for ripple and/or noise with a sensitive ac oscilloscope. Power-supply troubleshooting procedures were explained in Chapter 4.

Obscure trouble symptoms in high-sensitivity op-amp cir-

cuitry may be caused by leakage between conductors or terminals of the circuit board. If there is any suspicion of leakage, it is advisable to clean the circuit board carefully. A standard cleaning solvent should be used, after which the board may be dried rapidly in a stream of warm air. With reference to Fig. 10–12, the most probable cause of trouble in an op-amp differentiator is a defective capacitor. In other words, C may become leaky or open-circuited. If the capacitor is "open," normal operation will be resumed when it is "bridged" with an equivalent good capacitor. If C is leaky, the technician will usually measure more or less dc output voltage from the op amp. In other words, the dc voltage at the inverting input terminal and at the output terminal will be practically zero under normal operating conditions.

Note in passing that the configuration shown in Fig. 10–12 has been simplified for clarity. In application, two additional resistors are employed, as exemplified in Fig. 10–17. The function of R1 is to limit the high-frequency gain of the system. When high-frequency noise is a problem in differentiator operation, a low-pass RC filter may be utilized instead of the series resistor R1. Next, observe that R3 has the same value as R2. The function of R3 is to reduce the output offset voltage by making the bias currents flowing into each input approximately equal. In case C1 becomes leaky, this equality is upset, inasmuch as the input terminal is generally returned to ground through a comparatively low source resistance. On the other hand, if it is found that C1 is not leaky, an objectionable output offset voltage is probably being developed by a marginal op amp.

Next, consider the possible causes of malfunction in the op-amp integrating circuit depicted in Fig. 10–13. It is instructive to note that this simplified arrangement is impractical, because the low-frequency gain (dc gain) is excessive. In other words, there is no negative feedback with respect to dc. Accordingly, any small dc offset voltage developed at the output will be integrated; the arangement will ramp up or down until the op amp goes into saturation and becomes unworkable. Therefore, the configuration of Fig. 10–18 is employed in practice. Resistor R2 limits the low-frequency gain of the integrator, to minimize the possibility of ramping trouble symptoms. Note that at low frequencies, no integrator action occurs, and the arrangement operates simply as an inverting amplifier with a gain of 10 times in this example. Observe that resistor R3 acts to reduce output voltage, as explained previously. Thus, R3 also assists in preventing ramping trouble symptoms.

Stability is provided in the configuration of Fig. 10–18 by the

(a)

(b)

(c)

Figure 10–17. An op-amp differentiator with HF-limiting and output-offset balancing resistors. **(a)** Basic circuit configuration; **(b)** Equivalent offset-balancing circuit; **(c)** Appearance of a digital voltmeter. *(Courtesy of Dynascan)*

Figure 10–18. An op-amp integrator with limiting resistor, balancing resistor, compensating resistor, and compensating capacitor.

1.5 kΩ resistor and 100-pF capacitor. If this capacitor becomes open-circuited, the op amp is likely to oscillate and the system will have no integrating action. Note that the 100-pF capacitor is called a *compensating capacitor*. It acts to limit the high-frequency response of the op amp. Many op amps contain an internal compensating capacitor, and do not require external compensation. It is helpful to consider the need for compensation (phase compensation). Integrated transistors and resistors are associated with stray capacitance, and stray capacitances cause phase shift. This phase shift becomes greater at higher frequencies. In turn, if an op amp has very high frequency response, there will be some upper frequency at which the output voltage is shifted 180 deg in phase with respect to the input voltage. At this frequency, the intended negative feedback becomes positive feedback, and the op amp oscillates violently at this frequency. To prevent oscillation in Fig. 10–18, the 1.5 kΩ resistor and 100-pF capacitor are connected between designated terminals of the op amp, in order to reduce its internal high-frequency response.

As noted previously, the operational transconductance amplifier (OTA) is supplanting the OVA in some areas of the industrial-electronic field. The basic configuration for a typical OTA is depicted in Fig. 10–19. An amplifier bias current (ABC) terminal is provided, whereby the transconductance (gain) of the OTA can be controlled by an external bias current. Transistors Q1 and Q2 are connected in a current-mirror configuration. A current mirror serves to reverse the polarity of an input current. Thus, the amplifier bias current controls the gain of the differential pair Q3 and Q4 by varying their emitter bias. In turn, the differential output signal currents from Q3 and Q4

are applied to the second differential pair of transistors, Q7 and Q8. Note that Q10 and Q11 form another current mirror, which functions to transform the double-ended output of the transistor network Q5 through Q9 into a single-ended output. The output impedance is extremely high.

A gyrator is an op-amp arrangement that functions like an inductor by reversing the voltage-current relation of a reference capacitor. With reference to Fig. 10–20, the input 1 and input 2 terminals "look like" the terminals of a 10-kH inductor. In other words, when an input voltage is applied, the gyrator draws a lagging current. This phase reversal is accomplished by inverting the phase of the 3-Ωf capacitor via the op-amp configuration. Although an op-amp gyrator cannot handle substantial power, it finds valuable application in ac signal filters, such as high-pass, low-pass, and bandpass arrangements. The gyrator is very compact and replaces very large inductors in typical cases; such inductors would be objectionably bulky and costly in conventional form. Large inductors are also troublesome because it is difficult to prevent them from picking up stray magnetic fields. On the other hand, a gyrator is practically immune to stray-field pickup.

Troubleshooting a gyrator is comparatively simple, provided

Figure 10–19. Basic configuration of an operational transconductance amplifier.

Figure 10–20. An OTA gyrator arrangement.

that the OTA is socket-mounted. In other words, a replacement OTA can then be plugged into the socket to determine whether normal operation is resumed. However, if the OTA is soldered into its circuit board, systematic tests should be made to determine whether the OTA is faulty, or whether an external circuit component is defective. The reference capacitor falls under preliminary suspicion. To check whether the capacitor may be open, it may be temporarily "bridged" by a new capacitor. If normal operation is not resumed, capacitor leakage may be causing the malfunction. In most cases, the most practical approach is to cut a foil conductor to one end of the capacitor with a razor blade. In turn, an out-of-circuit check can be made for capacitor leakage. Gyrator malfunction can also be caused by off-value resistors, although this defect is infrequent. The technician should make certain that trouble symptoms are not being caused by contamination of the circuit board by foreign substances. Thus, the possibility of OTA faults is checked by elimination. After an OTA is replaced, it may be necessary to readjust the amplifier bias current to obtain normal device gain.

REVIEW QUESTIONS

1. Name some of the solid-state devices used in industrial-electronics equipment.
2. What are the chief characteristics of an op amp?
3. Give several applications for op amps in industrial electronics equipment.
4. Define an inverting input; a non-inverting input.
5. Distinguish between an OVA and an OTA.
6. Explain the basic principle of a differential amplifier.
7. Discuss the function of a constant-current source.
8. Why does a transistor with a fixed bias operate as a constant-current source?
9. If the voltage-source polarity to an op amp is accidentally reversed, what will be the result?
10. How can an op amp be protected against supply-voltage polarity reversal?
11. Is a gain of 1000 times unusual for an op amp?
12. Explain the nature of a virtual ground.
13. How is the system gain calculated for an op amp that employs substantial negative feedback?
14. Describe the circuit actions that result in precise differentiating action by an op-amp differentiator.
15. Explain the circuit actions that result in precise integrating action by an op-amp integrator.
16. How is a basic op-amp integrating circuit elaborated to prevent damage to the op amp?
17. Discuss summing action by an op-amp summer.
18. Explain subtracting action by an op-amp subtracter.
19. What is the meaning of the term "scaling," or "weighting"?
20. Describe op-amp logarithmic and anti-logarithmic amplifiers.
21. How are op amps configured to perform multiplying and dividing operations?
22. In preliminary trouble analysis, how are trouble symptoms localized to the op amp, or to the associated circuitry?
23. Explain some trouble symptoms that can result from leakage between conductors or terminals of a PC board.
24. How does a troubleshooter contend with an output-offset voltage trouble symptom?

Troubleshooting Solid-state
Television Receivers

11.1 GENERAL CONSIDERATIONS

A television receiver consists basically of a picture-signal processor and a sound-signal processor. The picture signal is associated with synchronizing signals, as depicted in Fig. 11–1. A sync section is provided in the receiver for processing these horizontal and vertical sync signals. Note that there are three basic signals generated within the receiver system: the RF local-oscillator signal, the vertical-deflection signal, and the horizontal-deflection signal. Thus, the technician is concerned with three types of ac voltages: the 60-Hz power source, the input signals to the receiver, and the signals that are generated within the receiver. As detailed subsequently, dc source voltage is obtained from rectification of the 60-Hz line voltage; dc voltage at the output of the picture detector is produced by rectification of the picture signal; AGC bias voltage also is produced by rectification of the picture signal; certain other dc base-bias voltages are produced by rectification of ac input signals; dc bias voltages at the bases of oscillator transistors result from rectification of signal voltages generated by the oscillators; dc bias voltage at the base of the converter results from rectification of the injected local-oscillator voltage; dc bias voltage at the base of the horizontal-output transistor is produced by rectification of the driving voltage; dc boost voltage results from rectification of the horizontal-deflection signal; and the picture-tube

high voltage results from rectification of the stepped-up horizontal-deflection signal waveform.

 With reference to Fig. 11–2, the three input signals are processed by the receiver circuits as follows: The channel selector tunes in a TV station signal. This picture-and-sound signal is then amplified and demodulated. Next, the sound signal is separated from the picture signal. After further amplification, the picture signal is applied to the picture tube, and the sound signal is fed to a speaker. A branch circuit from the picture amplifier separates the sync signal from the picture signal. The vertical-sync signal is then separated from the horizontal-sync signal, and these sync signals are fed to the deflection sections. Although the end result is always the same, various receivers

(a)

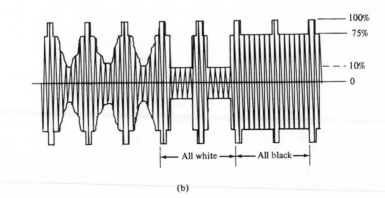

(b)

Figure 11–1. Basic TV transmitting and receiving system. **(a)** Picture-signal and sound-signal sections are utilized; **(b)** Synchronizing pulses accompany the picture signal.

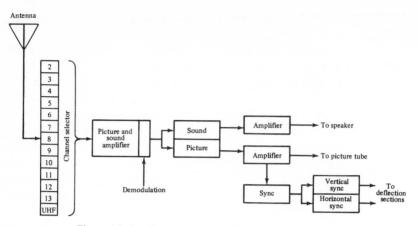

Figure 11–2. Arrangement of input-signal channels.

employ different methods of signal processing. In turn, the technician needs to be familiar with typical design features and variations.

11 . 2 BLOCK ANALYSIS

A TV receiver system can be broken down into block diagrams of sectional or subsectional form. A more elaborate block diagram provides greater utility in preliminary trouble analysis. As an illustration, Fig. 11–3 shows a generalized block diagram in which the sound section drives a speaker, and the picture section drives the picture tube, with the sync section branching off from the picture section. In turn, outputs from the sync section are applied to the horizontal-sweep and the vertical-sweep sections, which in turn drive the deflection yoke. From the standpoint of the troubleshooter, it is helpful to elaborate the sound and sync takeoff points, as shown in Fig. 11–4. Thus, if the video-output stage is "dead" in (a), there is no picture and no sound. However, if the video-output stage is "dead" in (b), the sound output continues, although there is no picture. In (c), the preliminary symptoms are the same as in (b); in (d), the preliminary symptoms are the same as in (a). Therefore, additional tests are required to close in on the faulty circuit in various cases. Chart 11–1 outlines the general troubleshooting procedure.

Observe in Fig. 11–4(a) that loss of both picture and sound does not necessarily throw suspicion on the video amplifier—the video detector might be at fault, for example. However, loss of both

picture and sound in (b) clears the video amplifier from suspicion, and directs the technician's attention to the video detector. Again, loss of both picture and sound in (c) clears the video amplifier from suspicion. On the other hand, loss of both picture and sound in (d) directs preliminary attention to the video amplifier. Note that when there is loss of both sound and picture, but a raster (brightness without a picture) is displayed on the screen of the picture tube, the power supply is cleared from suspicion. The trouble point could be located anywhere between the antenna and the sound-takeoff point. In this situation, the technician should observe the snow level on the screen (Fig. 11–5) when the contrast control is fully advanced. A high snow level indicates that the fault is in the RF section; on the other hand, a moderate snow level points to trouble in the IF section; again, a very low snow level or a snow-free raster throws suspicion on the video detector or the video amplifier.

Note in passing that when there is a dead stage in the video signal channel, the noise level from the speaker follows the same general rule as the snow level displayed on the picture-tube screen. In other words, when the volume control is fully advanced, there will be a loud random-noise output from the speaker if the trouble is in the RF stage, for example. However, if the trouble is in the video detector, the random-noise output will be much less. Noise analysis is not as informative as snow analysis, because of the limiting action provided by an FM detector. Observe that the absence of snow does not necessarily mean that the last IF stage or video detector is "dead." As pictured in Fig. 11–6, both the RF section and IF section are AGC-controlled. Accordingly, an AGC fault that biases off the IF section

Figure 11–3. A generalized block diagram.

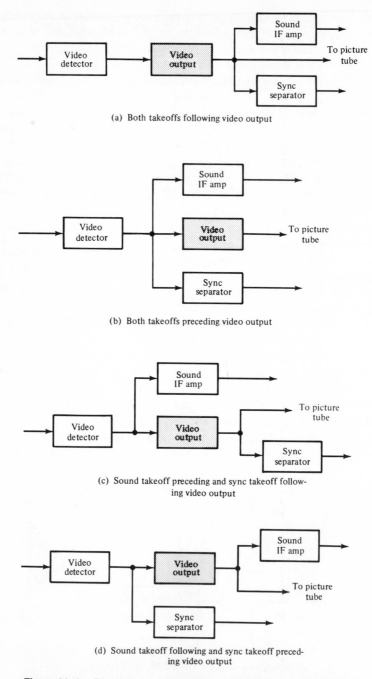

(a) Both takeoffs following video output

(b) Both takeoffs preceding video output

(c) Sound takeoff preceding and sync takeoff following video output

(d) Sound takeoff following and sync takeoff preceding video output

Figure 11–4. Block diagram showing sound and sync takeoff points.

CHART 11-1

General Troubleshooting Procedure

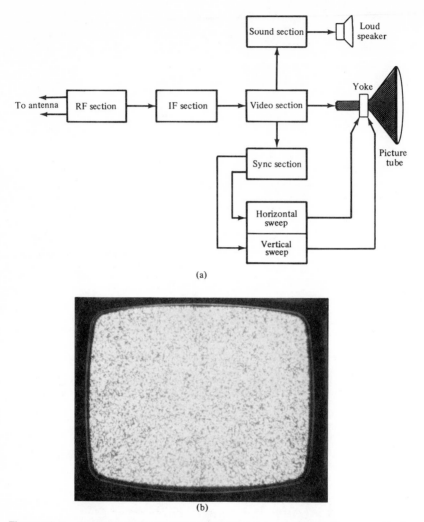

(a)

(b)

Figure 11–5. Preliminary localization is facilitated by observing the snow level. **(a)** Elementary block diagram; **(b)** Example of high snow level.

can also be responsible for a no-snow display. This topic will be detailed subsequently.

A troubleshooter considers the antenna and lead-in to be part of the RF section, because a weak- or no-picture symptom could be caused by failure of the antenna installation. Poor connections in the antenna system can cause flashes, a flickering picture, or intermittent reception. Note that an antenna fault that prevents picture production may nevertheless permit sound reproduction. In other words, although

the picture signal becomes so weak that no picture is displayed, the sound generally remains sufficiently strong that audio output continues, although the reproduction may be noisy. In remote areas where the signal strength is weak, preliminary trouble analysis is facilitated by the use of a test-pattern generator (TV analyzer) such as that illustrated in Fig. 11–7.

(a)

(b)

Figure 11–6. A defective AGC section can bias off the IF section. **(a)** Relation of AGC section to the RF and IF sections; **(b)** Transistor transconductance variation versus base-emitter (AGC) voltage.

(a)

(b)

Figure 11–7. A test-pattern generator provides a TV signal with adjustable level. **(a)** Appearance of instrument; **(b)** Test pattern display. *(Courtesy of Dynascan)*

11.3 TUNER (FRONT END) SECTION

A TV tuner section contains three stages or subsections, as depicted in Fig. 11–8. These are termed the RF amplifier, oscillator, and mixer circuits. A malfunction in the oscillator stage that causes it to oscillate at an incorrect frequency may cause the channel-4 picture to be received on the channel-3 setting of the tuner, for example. Note that a defect in the RF stage results in a weak and snowy picture, or no picture. Again, if an oscillator fault causes the RF amplifier to receive the channel-4 signal on the channel-3 setting, the picture will be weak, because the RF circuits are operating off-resonance. Again, if the oscillator ceases operation, there will be no picture or sound reproduction, and the raster will be filled with snow. The mixer stage

(a)

(b)

Figure 11–8. The front end or tuner contains three subsections. **(a)** Block diagram; **(b)** Appearance of a VHF and UHF tuner.

is the strongest source of random noise, or snow. If the mixer stage becomes defective, there will be a considerable drop in the snow level displayed on the picture-tube screen. Beginners should note that in a weak-signal area, pictures are always snowy, because the AGC section then permits the RF and IF sections to operate at maximum gain.

With reference to Fig. 11–9, dc bias voltages at the bases of both the oscillator and mixer transistors are obtained, at least in part, from the oscillator signal. If the technician can measure the voltage at the base of the oscillator transistor (at C127 in Fig. 11–9), he then has a definite check on oscillator operation. Similarly, if the dc injection voltage can be measured at the base of the mixer transistor,

Figure 11-9. Typical TV tuner configuration.

a definite check is provided concerning whether the oscillator output is being injected into the mixer transistor, as it should be for normal heterodyne action. Observe that Q3 is reverse-biased in normal operation of the oscillator circuit. This is the result of class-C oscillator operation, wherein the collector current flows in short pulses, and considerable base current is drawn. This base current is also drawn in short pulses, and their rectified average value makes the base more positive than the emitter, as measured by a dc voltmeter.

11 . 4 IF AMPLIFIER TROUBLESHOOTING

Referring to Fig. 11–10, note that a completely "dead" stage will block passage of picture, sound, and sync signals. A useful quick check can be made with a 500-pF capacitor. As an illustration, if the technician suspects that the second IF stage is "dead," the capacitor can be connected temporarily between the base and collector terminals of Q2 thereby bypassing the signal around Q2. If a weak picture is then displayed (and sound is reproduced), it is concluded that Q2 is open, shorted, or biased off, or that it lacks collector voltage. Again, the bridging or jumping capacitor can be connected temporarily from the collector of Q1 to the base of Q2. If the receiver resumes operation, it is indicated that T1 is defective (for example, the secondary winding may be open). Similarly, the bridging capacitor can be temporarily connected from the collector of Q2 to the base of Q3. If picture and sound reproduction resume, it is concluded that T3 is defective.

Note that when a transistor is defective, it will not respond to a signal injected from a generator, or from a TV analyzer such as that illustrated in Fig. 11–7. Injection tests should be made with the AGC line clamped to normal operating voltage by means of a bias supply. This procedure avoids the possibility of confusing AGC trouble with IF-amplifier trouble. To close in on defective components, dc-voltage measurements are usually adequate. Measured values are compared with specified values in the receiver service data. Transformer continuity checks can be made with an ohmmeter, since IF coils are often shunted by resistors. Such resistors, however, have much higher values than winding resistance values. Note carefully that an open neutralizing capacitor [C in Fig. 11–10(b)] is likely to result in stage oscillation, thereby blocking signal passage. In such a

Figure 11–10. IF amplifier circuitry. **(a)** Unneutralized IF stages; **(b)** A neutralized IF stage.

case, a TVM connected at the output of the picture detector will indicate a large dc voltage, such as 5 or 10 volts.

Oscillation can also be caused by misalignment in some cases. For example, if all of the IF tuned circuits are accidentally peaked to the same frequency, the system is likely to break into uncontrolled oscillation. Note that realignment of tuned circuits is ordinarily done

Figure 11-11. A TV sweep-alignment and marker generator. *(Courtesy of Sencore, Inc.)*

after all troubleshooting is completed, unless it is known or suspected that the alignment adjustments have been tampered with. Alignment of IF (or RF) circuits is considerably facilitated by the use of a sweep-alignment and marker generator, such as illustrated in Fig. 11-11. The service data for the particular receiver should be referred to for detailed instructions of alignment procedure.

11 . 5 VIDEO-AMPLIFIER TROUBLESHOOTING

It is evident that the input terminal of the picture detector is included in the IF system, but that the output of the picture detector is part of the video-amplifier system. When a detector diode is replaced, it is essential to observe correct polarity. Otherwise, a negative picture will be displayed, as illustrated in Fig. 11-12. Note that the normal voltage gain of a video amplifier is approximately 100 times, with the contrast control set to its maximum position. Since Q204 operates as an emitter follower, it provides no voltage gain, although it de-

(a)

(b)

Figure 11–12. Basic video-amplifier arrangement. **(a)** Partial schematic diagram; **(b)** Negative picture display. *(Courtesy of Dynascan)*

velops considerable current gain with impedance matching from the picture detector to the base of Q205. Thus, all of the voltage gain is provided by Q205. A technician checks gain with an oscilloscope, using a low-capacitance probe, and using a test-pattern signal. From a general viewpoint, troubleshooting a video amplifier is essentially the

same as troubleshooting an audio amplifier. The chief difference is in the much greater bandwidth of a video amplifier (4 MHz versus 20 kHz).

Video-amplifier trouble should not be confused with picture-tube trouble. If a scope test shows a 50-volt p-p signal at the cathode of the picture tube, and the dc voltages at the picture-tube terminals are normal, it is logical to conclude that the video amplifier is operating normally, and that the trouble is in the picture tube. Note that a TV analyzer such as that illustrated in Fig. 11–7 provides a video-frequency output signal that can be injected at the base of the video driver transistor, at the base of the video-output transistor, or at the cathode of the picture tube. Most of the trouble symptoms encountered in video amplifiers are caused by leaky or open coupling, bypass, or decoupling capacitors. Since the video-output transistor operates at a comparatively high collector voltage, and dissipates appreciable power, it is also a ready suspect.

11 . 6 AGC SECTION TROUBLESHOOTING

A typical AGC system for a solid-state TV receiver is shown in Fig. 11–13. It consists of a closed loop comprising the AGC gate, the RF and IF amplifiers, video detector, and the first video amplifier. Trouble-shooting of the AGC system involves dc-voltage measurements under signal and no-signal conditions, combined with a good understanding of system operation. Note that the AGC section maintains a relatively constant output of 1.2 volts at the emitter of the first video amplifier; this output normally varies only a small amount over a wide range of signal-input voltage. Nevertheless, the small variation that occurs provides a wide range of control action, as detailed below. This is an example of a keyed or gated AGC system in which the AGC control voltage is developed during the horizontal-sync interval, and sustained by the AGC circuitry for the duration of the forward-scanning interval.

Note that as the input signal to the RF amplifier increases, the output of the first video amplifier tends to increase accordingly. However, any increase in video signal level is applied as an input to the AGC gate. A 30-volt negative pulse opens the AGC gate during flyback time, during which the sync pulse component of the video signal is amplified. In turn, a positive AGC voltage is developed. To prevent the collector-base junction of Q501 from becoming forward-biased by the AGC voltage, CR501 is connected in series with the collector circuit. The AGC bias voltage is applied as forward bias

Figure 11–13. Typical AGC system for a solid-state TV receiver. **(a)** Block diagram; **(b)** Simplified schematic. *(Adapted from an RCA original)*

to the RF-amplifier transistor, which in turn operates nearer saturation at reduced gain. Observe that the RF-amplifier transistor amplifies the incoming RF signal, and also functions as a dc amplifier for the AGC voltage. In turn, the AGC voltage is applied in opposite polarity to the base of the first IF amplifier. This IF-amplifier transistor is controlled by reverse bias.

To maintain optimum signal/noise ratio, delayed AGC is employed, whereby the RF amplifier operates at practically full gain on weak-to-medium signals, with most of the AGC control provided by the IF section. On strong signals, the RF amplifier operates at progressively reduced gain. A total AGC range of 70 dB is available. Note that the point at which the RF transistor starts to operate at reduced gain is determined by the setting of the bias-delay-adjustment control R112 in Fig. 11–13. In other words, this control establishes the minimum gain of the first IF amplifier. Filtering of the AGC voltage is provided by C556, C111, R508, and R509. Additional filtering is provided by R113 and C202. As a practical note, most trouble symptoms in AGC systems are caused by leaky filter capacitors. However, if all of the capacitors are in good condition, the gating and delay diodes should be checked for front-to-back ratio. In some cases, the AGC winding on the flyback transformer becomes open-circuited, or leakage develops between the winding and the transformer core.

11.7 TROUBLESHOOTING THE
VERTICAL-DEFLECTION SECTION

A typical vertical-deflection system comprises an oscillator, predriver, driver, and a vertical-output stage, as depicted in Fig. 11–14. This free-running waveform generator is synchronized by the incoming vertical-sync pulses. A feedback circuit is employed to provide oscillation; positive feedback occurs from the yoke to the base of the oscillator transistor Q504. Observe that this feedback voltage is coupled through R534 and C513. A defective component in the feedback loop, such as a leaky coupling capacitor, can cause puzzling trouble symptoms. The technician may regard the oscillator transistor as a switch that is conducting during the retrace interval, and nonconducting during the downward-scanning interval. Note that spurious oscillation is prevented by the feedback action of C515. In other words, if the oscillator operates erratically, C515 falls under preliminary suspicion.

During the downward-scanning interval (ramp portion of the

Figure 11–14. Typical vertical-deflection arrangement. **(a)** Block diagram; **(b)** Basic schematic diagram. *(Adapted from an RCA original)*

sawtooth waveform), Q504 is nonconducting for a period that is determined by the time-constant of C513, R251, and the vertical-hold control R253. Conduction starts in Q504 when its base voltage becomes lower than its emitter voltage. Retrace is rapid, and feedback from the emitter circuit of Q103 ensures precise starting of conduction. Note that the sweep capacitor C516 discharges rapidly during retrace owing to the 130-volt flyback pulse. Next, during downward scan, the capacitor voltage builds up gradually. An S-correction filter is included to modify the scanning voltage slightly at

the beginning and end of the ramp, to obtain a precisely linear picture display.

As a practical note, CR503 and C541 comprise an antihunt network to prevent variation of alternate cycles of the scanning waveform. In other words, if a trouble symptom occurs in which two different sizes of rasters are superimposed, the antihunt components fall under preliminary suspicion. The voltage-dependent resistor (VDR) RV101 is provided from collector to emitter of the vertical-output transistor to protect the transistor from damage by the high-voltage flyback pulse. Note also that a varistor RV501 is included to hold the bias voltage of the oscillator transistor constant. In receivers that have been in extended use, trouble symptoms are sometimes caused by worn and "noisy" operating controls, such as the vertical-hold control. From the standpoint of probability, a vertical-output transistor is more likely to fail than an oscillator transistor, inasmuch as the former operates at a much higher power level.

11.8 HORIZONTAL SWEEP CIRCUIT TROUBLESHOOTING

As depicted in Fig. 11–15, a horizontal-deflection system provides deflection of the electron beam in the picture tube, high voltage for the second anode of the picture tube, and boost voltage for the accelerating anode of the picture tube. Other functions include horizontal blanking, automatic frequency control, and fail-safe functions. Observe that horizontal sync pulses from the sync separator enter the phase splitter. In turn, the phase splitter develops two output voltages, each consisting of equally amplified horizontal-sync pulses that are oppositely polarized. These pulses are compared in frequency and phase with the horizontal-sweep voltage. If there is a discrepancy in frequency or phase, the AFC circuit produces an error voltage that brings the horizontal oscillator back in step with the horizontal-sync pulses.

With reference to Fig. 11–16, the horizontal oscillator provides a 15,750-Hz square-wave voltage. A blocking-oscillator arrangement is utilized. Bias stabilization is provided by thermistor RT501. Note that the horizontal-driver stage amplifies the output from the oscillator and shapes it into a pulse with a width of 18 microseconds. This pulse drives the horizontal-output transistor, shown in Fig. 11–17. Observe that the high-voltage (flyback) transformer, yoke, and damper (CR105) operate as an emitter follower load for Q102. As

Figure 11–15. Typical horizontal-deflection system. **(a)** Block diagram of AFC and sweep system; **(b)** Arrangement of AFC section. *(Adapted from an RCA original)*

shown in Fig. 11–18, the output transistor goes into conduction at the center of the forward-scan interval, and produces a linear ramp for the last half of the scan. Then, at the end of the scan, Q102 is abruptly turned off by the driver waveform, and a high-amplitude "flyback pulse" is generated. This is the flyback or retrace interval that returns the electron beam to the left-hand edge of the picture-tube screen and causes damper CR105 to conduct. Note that the damper conducts for the first half of the forward-scan interval, after which it goes into nonconduction.

Means are provided to protect the horizontal-output transistor from excessive current flow by inclusion of the current-limiting transistor Q108. In turn, CR107 bypasses spike voltages that could damage Q108. High voltage is developed by stepping up the flyback pulse in T102. A 13-kV pulse results, which is rectified and fed to the

Figure 11-16. Basic configuration for a horizontal-oscillator and driver arrangement. *(Adapted from an RCA original)*

picture tube. Also, the flyback pulse is rectified by CR508 and filtered to supply +240 volts to the accelerating anode of the picture tube. Note that the horizontal-output transformer also provides a comparison pulse for the AFC section, and an AGC keying pulse. Picture width is controlled by adjustment of L107, which also affects the efficiency of sweep operation. It is usually adjusted on the basis of current flow as indicated by a milliammeter.

Various picture symptoms result from different defects in the horizontal-output and high-voltage sections. Some of these symptoms provide definite localization, and others indicate only a probability or possibility of trouble in certain sections or circuits. In turn, localization tests are often required. The chief test instruments employed in these tests are the high-voltage dc voltmeter, oscilloscope, pattern generator, and specialized instruments, such as a TV analyzer. A typical high-voltage probe with indicating meter is illustrated in Fig. 11–19. As explained in greater detail below, a TV analyzer provides a pulse output that can be used to drive a horizontal-output system. Sync-pulse and AGC keying-pulse outputs are also available. A TV analyzer also includes a flyback-transformer and yoke tester, to indicate the presence of shorted turns.

Picture trouble symptoms include an absence of raster (dark screen), inadequate raster width (narrow picture), excessive height and width with reduced brightness (blooming), cramped image at the left, right, or in the center region (horizontal nonlinearity), one or more bright stripes with associated overlap of the image (horizontal

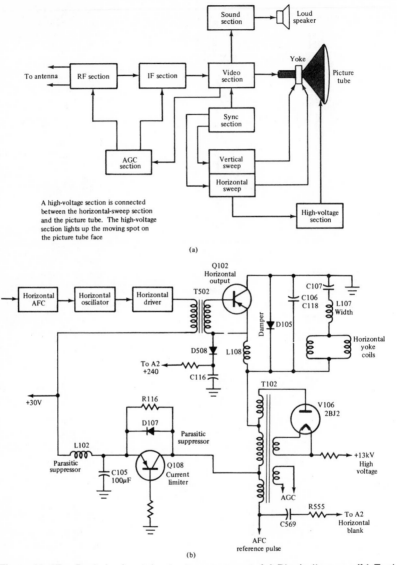

Figure 11–17. Basic horizontal-output arrangement. **(a)** Block diagram; **(b)** Typical configuration of a horizontal-output section. *(Adapted from an RCA original)*

foldover), trapezoidal-shaped raster (keystoning), impaired AGC action with resulting abnormal or subnormal picture contrast, and defective horizontal-sync lock. In most cases, a dark-screen symptom results from complete or partial failure of the high-voltage supply. However, the technician should not disregard the alternate possibilities of a

Figure 11–18. Circuit-action sequence in the horizontal-output system. *(Courtesy of RCA)*

Figure 11–19. A high-voltage probe with indicating meter. *(Courtesy of Dyna-scan)*

defective picture tube, or incorrect dc voltages at the cathode or grid of the picture tube. Therefore, the first step is to measure the high-voltage value at the accelerating anode of the picture tube. (See Chart 11–2.)

High-voltage failure can result from component defects in either the high-voltage section or in the horizontal-deflection section. As an illustration, a semiconductor rectifier stack used for rectifying the high-voltage output can become open-circuited or short-circuited. In such a case, there is no high-voltage output, although the high-voltage ac input is normal. If a high-voltage capacitance-divider probe is available, the ac input voltage can be measured with a TVM or an oscilloscope. However, capacitance-divider probes are generally regarded as laboratory equipment, and are not in wide use by service technicians. Accordingly, a substitution test is usually made when rectifier failure is suspected. It is evident that high-voltage failure can also be caused indirectly by drive failure from the horizontal oscillator. In turn, a basic test consists of a waveform check at the input of the horizontal-deflection section. The drive waveform should have the approximate amplitude specified in the receiver service data, and should also have the correct waveshape. One of its most im-

CHART 11-2

Troubleshooting Guide For A No-raster Symptom

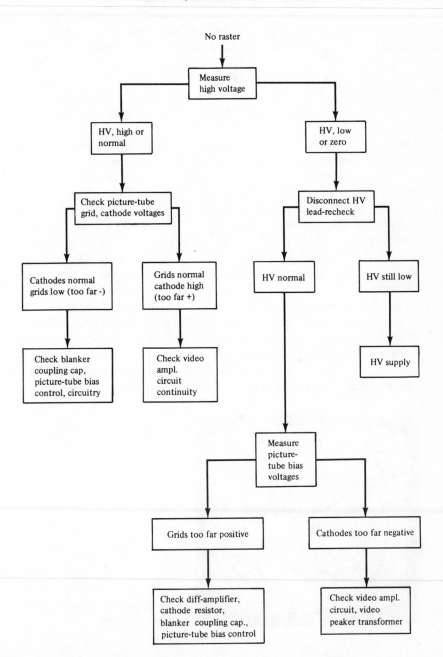

portant characteristics is rapid rise and fall, so that the switching action of the horizontal-output transistor occurs with great rapidity.

Many technicians prefer to make signal-substitution tests instead of waveform checks. As an illustration, when failure of drive voltage is suspected, a suitable drive substitution voltage can be applied from a TV analyzer. Then, if a raster is displayed on the picture-tube screen, the suspicion of drive failure is confirmed. Another approach in this situation consists in the substitution of a high voltage to the picture tube. For example, a TV analyzer may provide a 15-kV dc output, which can be applied to the accelerating anode of a picture tube. If the raster reappears when the high voltage is applied, it is logical to conclude that a defect is present in the high-voltage circuitry. A narrow-picture symptom, as illustrated in Fig. 11–20, can be caused by defects in sections other than the horizontal-output circuit. Thus, low-supply voltage often becomes first evident as subnormal picture (raster) width. Again, attenuated or distorted drive waveforms can also cause this picture symptom. Therefore, these possibilities should be checked at the outset.

If the trouble is localized to the horizontal-output section, the defective component may be in the damper circuitry or in the output circuitry. It follows from the yoke-current waveform depicted in Fig. 11–18 that reduction in width (nonlinearity) in the left-hand portion of the image is likely to be caused by damper circuit defects. On the other hand, if the cramping or nonlinearity occurs in the right-hand portion of the image, the trouble is likely to be found in the output circuitry. *Blooming* is a technical term for an enlarged and dim picture that is expanded both horizontally and vertically with the outer portion

Figure 11–20. A "narrow-picture" trouble symptom.

Figure 11–21. Example of picture blooming.

of the image off-screen, as exemplified in Fig. 11–21. This picture symptom almost invariably indicates a subnormal value of high-voltage supply to the picture tube. Note that the high-voltage value should be made under load, as the regulation of the supply may be very poor under fault conditions. The most common cause of blooming is a defective high-voltage filter capacitor.

11 . 9 TV ALIGNMENT

A sweep-frequency generator and oscilloscope are employed in the alignment of TV receivers. Either a separate marker generator or a built-in marker generator is also required. A post-marker/sweep generator is illustrated in Fig. 11–22. The advantage of post-marking is that there is no possibility of distorting a response-curve display from marker-signal overload. Several precautions should be observed in TV alignment procedure, as follows:

1. Disable the high-voltage supply to the picture tube by removing the high-voltage lead to the picture tube. Then connect the load specified by the receiver manufacturer from B+ to ground. Thereby, a proper load is provided for the low-voltage power supply. Do not permit the disconnected high-voltage termination to touch the receiver chassis.
2. Disable the horizontal-oscillator transistor to prevent horizontal pulses from interfering with the response-curve display.

Figure 11–22. A post-marker/sweep generator for TV alignment. *(Courtesy of Heath Company)*

3. Disable the vertical-oscillator transistor to prevent vertical pulses from interfering with the response-curve display.

4. Use the correct cables supplied with the sweep-alignment equipment.

5. Do not place the alignment generator or oscilloscope leads over an IF board, IF coils, or devices, because of possible disturbance or oscillation of the circuits under test.

6. Check and make certain that there are no ground loops between the units of equipment. This possibility is checked by observing the scope pattern and touching each item of equipment. If the trace moves or changes shape, check all ground connections. Sometimes it is necessary to ground all of the cables at one common point.

7. Bias voltage(s) as specified by the receiver manufacturer should be applied to the receiver under test.

8. Increased accuracy in trap adjustments can be obtained by reducing the bias on the IF stages. In such a case, be careful not to overload the IF strip by excessive generator signal amplitude.

9. Set the vertical-gain control of the oscilloscope near its most sensitive position and keep the generator output as low as possible. This precaution will avoid overloading

the IF strip with resulting curve distortion and mis-
alignment.

10. When one is adjusting a coil or transformer that has two
 slugs, each slug may have two positions which appear to
 tune to the correct frequency. However, the position
 farthest away from the center of the coil is the correct
 position. The position nearest the center of the coil is in-
 correct owing to coupling from one coil to the other
 through the slugs. It may be advisable to turn both slugs
 out to the ends of the coil form, and then turn the slugs,
 each slug a little at a time, to the correct positions.

11. TV receiver alignment makes greater demands on the skill
 and understanding of the technician than does any other
 service operation. Therefore, it is essential for the tech-
 nician to recognize the symptoms of misalignment. The
 order in which the various sections of a receiver are
 aligned may differ between receivers of the same manu-
 facturer. Accordingly, the alignment procedure for the
 specific receiver should be followed in detail.

The receiver manufacturer's instructions should be followed
exactly. If the alignment instructions do not indicate a specific setting
for the channel selector, turn the selector control to the highest un-
used VHF channel. In very strong signal areas, it may be necessary to
disable the VHF oscillator to prevent TV station interference during
alignment procedures. Alignment should be made at the same approxi-
mate temperature at which the receiver will be viewed. Referring to
Fig. 11–23, proceed as follows:

1. Connect the generator, attenuator, and oscilloscope to-
 gether, and connect the generator and attenuator to the
 TV receiver as indicated in the diagram. In case bias volt-
 age is called for, connect the bias leads at the proper points
 in the receiver. Connect a dc voltmeter across the bias bind-
 ing posts and adjust the voltage to the value specified by
 the receiver manufacturer. Either positive or negative bias
 may be applied by the corresponding setting of the bias
 switch.

2. Place both of the 20-dB rocker switches to the "in" position
 for 40 dB of attenuation.

3. Set the oscilloscope vertical attenuator and vertical-gain
 control to near maximum sensitivity.

Figure 11–23. Connections for IF sweep and trap adjustment. *(Courtesy of Heath Company)*

4. Turn the sweep-range switch to its IF position. Turn on the TV receiver and allow it to reach normal operating temperature. This trace shown in Fig. 11–24(a) is a typical over-all IF response curve found in many TV receivers. At the start, the curve shape may be abnormal. However, the markers on the display should follow a 42.17, 42.75, 45.00, and 45.75 MHz sequence. If not, throw the "trace-reverse" switch to its opposite position.

5. Turn on the 45.00-MHz marker and place the "blanking" switch in its "off" position.

6. Adjust the "sweep center" control and locate this marker.

7. Adjust the "phase" control and superimpose both markers; then throw the "blanking" switch to its "on" position.

8. Turn on the required switches. If the TV set's IF circuits are correctly aligned, the marker should appear as shown

in Fig. 11–24(a). If the IF circuits are far out of alignment, you may have to inject the signal into the last IF stage. Then work back toward the IF input. The output from the generator ("trace" and "marker" controls) should be increased when the signal is injected into the last IF stage, because there will be very little IF amplification. In turn, the generator output *must* be decreased as the signal-injection point is moved back toward the input of the IF circuits. An over-all IF check and adjustment should be performed after the stage-by-stage adjustments have been made.

Figures 11–24 through 11–26 show normal and incorrect IF curves. Figure 11–24(a) depicts a normal IF curve with proper size markers. Figure 11–24(b) exemplifies a normal IF curve with excessive marker amplitude. A distorted curve is shown in Fig. 11–25; in this example, the generator phase control is incorrectly adjusted. When the trace control is turned fully counterclockwise, only markers appear on the trace, as shown in Fig. 11–26. Some receiver manufacturers specify the positions of marker frequencies as so many "dB down," or, as a percentage of full trace amplitude. Figure 11–27 exemplifies an IF response curve with markers indicated in terms of both dB and percentage. The IF bandwidth is usually specified as the level at which the response is 6 dB (50 percent) down from maximum.

Proper bandwidth is necessary for reproduction of adequate picture detail. Subnormal bandwidth causes picture smear and loss of detail. Excessive bandwidth causes a grainy herringbone picture with possible ringing and the possibility of adjacent-channel interference. It is good practice to align a TV receiver for the bandwidth recommended in the pertinent service manual. Next, trap alignment

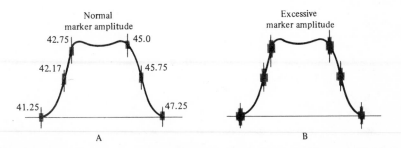

Figure 11–24. Typical overall IF frequency-response curves. *(Courtesy of Heath Company)*

Incorrect phase adjustment

Figure 11–25. Example of a distorted IF response curve. *(Courtesy of Heath Company)*

includes adjustments of the sound trap and adjacent-channel traps for separation of the video and sound signals. The sound and the adjacent-channel traps are adjusted to the frequencies specified by the receiver manufacturer. When aligning traps adjust the AGC bias according to the service instructions. The modulated RF output from the generator should be kept as low as possible and still produce an adequate pattern on the scope screen. Proceed as follows:

1. Turn the generator "sweep range" switch to its "off" position.
2. Turn on the 41.25-MHz marker switch and the "modulation" switch.
3. Adjust the sound trap for a minimum amplitude of the scope pattern. If necessary, reduce the IF bias to increase the IF gain. As the IF gain is increased, a more critical adjustment of the trap response can be made.
4. Again adjust the sound trap for minimum amplitude of the scope pattern.
5. Repeat the preceding three steps until the sound trap is properly aligned. Then turn off the 41.25-MHz marker switch.
6. Turn on the 47.25-MHz marker. Then follow the same procedure to properly align the adjacent-channel sound trap. If the receiver has a 39.75-MHz adjacent-picture trap, align the trap in accordance with the service instructions.
7. Recheck the over-all IF alignment for proper marker placement and bandwidth.

Trace gain off

Figure 11–26. Marker display with sweep signal reduced to minimum. *(Courtesy of Heath Company)*

Figure 11–27. Marker levels indicated in dB and percentage of maximum amplitude. *(Courtesy of Heath Company)*

REVIEW QUESTIONS

1. Why does the television troubleshooter need to be familiar with typical design features and variations?
2. Explain the significance of the following trouble symptoms: No picture and no sound; no picture with sound normal; no sound with picture normal.
3. Does the loss of both picture and sound reproduction necessarily throw suspicion upon the video amplifier?
4. What does a high snow level indicate; a moderate snow level; a low snow level?
5. Discuss noise-level outputs from the speaker in various trouble situations.
6. How does a test-pattern generator facilitate preliminary symptom analysis?
7. Name the three subsections of a VHF tuner.
8. Explain how a local oscillator is checked by dc-voltage measurement.
9. Why does a class-C oscillator often appear to operate with a reverse-biased transistor?
10. How is an IF signal temporarily fed through a suspected "dead" IF stage with a 500-pf capacitor?
11. Describe how an AGC line is clamped.
12. Explain the effect of IF oscillation on the picture-detector output voltage.
13. Can IF oscillation by caused by misalignment?
14. What picture symptom results if a replacement picture-detector diode is accidentally connected in reverse polarity?

15. How does a technician check signal gain in a video-amplifier stage?

16. State the component defects that are most likely to cause video-amplifier malfunction.

17. Is a total AGC range of 70 dB typical?

18. Discuss the function of a vertical antihunt circuit.

19. How would you proceed to analyze the following trouble symptoms: Dark screen; narrow picture; blooming; horizontal nonlinearity; horizontal foldover; keystoning?

20. Name the instruments that are used in alignment of TV receiver tuned circuits.

21. What are "ground loops" in an alignment test setup; how are they detected and corrected?

22. Explain the "pitfall" of overloading during alignment procedures.

23. Discuss correct and incorrect slug positioning in tuned transformers.

24. Would you check the alignment of a TV receiver first or last in your troubleshooting procedure?

Chapter Twelve

Troubleshooting Solid-state
Digital Equipment

12.1 GENERAL CONSIDERATIONS

Electronic digital computers employ various switching arrangements
to perform mathematical operations such as addition, subtraction,
multiplication, and division. Transistors, diodes, and integrated cir-
cuits are used in computer systems; nearly all of the larger computer
units utilize integrated circuitry. Many millions of basic arithmetical
operations can be performed per second by modern digital computers.
A typical electronic data processing system is illustrated in Fig. 12–1.
Integrated circuits are commonly mounted on modules, as exemplified
in Fig. 12–2. Modular construction minimizes down time, because the
technician need only localize trouble symptoms to a particular module,
which is then quickly replaced. In turn, the faulty module can be re-
paired at some convenient future time, or discarded according to the
judgment of the troubleshooter.

 Although the configuration of a digital computer appears to be
enormously complex, it will be recognized that comparatively simple
circuits are used to perform specified electronic switching functions.
These "building blocks" are arranged in series for sequential operation,
or are arranged in parallel for simultaneous operation. These facts
enable the troubleshooter to analyze digital circuitry with compara-
tively simple test equipment. To anticipate subsequent discussion, the
three chief "building blocks" of a digital computer are the *flip-flop*,
the *gate*, and the *pulse generator*. A flip-flop (FF) is a bistable multi-

Figure 12–1. A typical electronic data processing system. *(Courtesy of Hewlett-Packard)*

vibrator configuration that is applied typically to store digital-logic facts. A gate is a circuit having two or more inputs and one output, the output depending upon the combination of digital-logic signals applied to the inputs. It is typically utilized to determine a logical conclusion. A pulse generator is a free-running multivibrator, used to command flip-flop operation. In other words, the pulse generator, or clock, controls the sequence of switching operations in the computer system.

Consider the operation of the flip-flop illustrated in Fig. 12–3; its schematic diagram is depicted in Fig. 12–4. One transistor will be in the conducting state while the other transistor is in the nonconducting state. In other words, if Q1 is conducting, Q2 is nonconducting, and vice versa. Suppose that Q1 is conducting and Q2 is nonconducting. In turn, CR1 is forward-biased and CR2 is reverse-biased. Q3 is forward-biased and lamp I1 glows. CR1 and CR2 are called *steering diodes*. The steering diodes prevent positive pulses from triggering the flip-flop. To make the lamp stop glowing, terminals A and B can be momentarily connected, or a negative trigger input can be applied.

Figure 12–2. An integrated-circuit module.

Again to make the lamp glow, terminals D and E can be momentarily connected, or a negative trigger input can be applied. Note that when A and B are momentarily connected, Q1 is cut off and Q2 is biased into conduction. Thus, the state of the flip-flop is reversed. If a negative trigger input is applied, Q2 is driven into conduction, whereupon Q1 is driven into cutoff. The flip-flop will remain in either one of its states indefinitely until another negative trigger pulse is applied. The trigger pulse should have an amplitude of 5 or 6 volts in this example.

Flip-flops can serve as "memories," add or subtract pulses sequentially, count, and control other switching circuits. With refer-

Figure 12–3. A discrete-component flip-flop with an indicator light. *(Courtesy of Libe Company)*

Figure 12–4. Configuration of a typical flip-flop and indicator-lamp arrangement.

ence to Fig. 12–5, flip-flops may be connected in a chain for binary storage of digital pulses. This chain arrangement is called a *counter*. Computers employ the binary system of numbers, which is based on powers of 2 instead of powers of 10. As an illustration, in the decimal system, the number 397 can be expressed in the form $(3 \times 10^2) + (9 \times 10^1) + (7 \times 10^0)$. Observe that the position of a digit indicates how many times this power of 10 is to be added to itself. For instance, in the number 30, the position of 3 corresponds to the first power of 10, and the value of 3 indicates that $10(10^0)$ is to be added to itself three times.

In the binary system, all numbers are expressed in terms of the digits 0 and 1. With reference to a flip-flop, 0 and 1 correspond to the *off* and *on* states of a transistor. As an illustration, the sequence of flip-flop states *on-off-on* or 101 in binary notation, can be expressed as $(1 \times 2^2) + (0 \times 2^1) + (1 \times 2^0)$, which is equal to 5 in decimal notation. Observe in Fig. 12–5 how the decimal numbers up to 5 are counted in binary form by a flip-flop chain. In other words, the input trigger pulses are stored in the chain progressively as they are counted, and indicating lamps may be used for readout of the binary numbers. Note that the signal flow is depicted as progressing

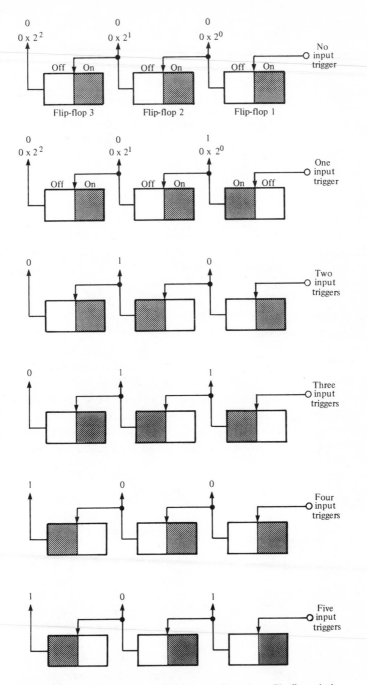

Figure 12-5. Development of binary numbers by a flip-flop chain.

(a) OR gate (b) NOR gate

Figure 12–6. Typical single transistor OR and NOR gates.

from right to left in Fig. 12–5, so that the binary numbers will be indicated in conventional order. After five input trigger pulses have been applied, the state of the counter corresponds to the binary number 101. If it is desired to reset the counter back to 000 at any time, a pulse is applied to the first transistor in *each* flip-flop, thereby driving the transistors into conduction.

Next, consider the basic logic gates, called AND, OR, NAND, and NOR gates. Logic gates are used with flip-flops in practically all computer circuitry. Transistor OR and NOR circuits are depicted in Fig. 12–6. An OR gate produces an output pulse when a positive pulse is applied at either of the input terminals. In other words, a positive pulse can be applied to R1, and an output is obtained. Similarly, a positive pulse can be applied to R2 and an output pulse is

(a) AND gate (b) NAND gate

Figure 12–7. Typical single transistor AND and NAND gates.

obtained. Again, if positive pulses are applied simultaneously to both R1 and R2, an output pulse is obtained. A CB configuration is employed, and the output pulse is positive-going. On the other hand, if a CE configuration is utilized, as depicted in Fig. 12–6(b), a positive-going output pulse is obtained when a negative-going input pulse is applied. In turn, the configuration is called a NOR (NOT OR) gate, because it reverses the input pulse polarity. Observe that the transistor in the NOR gate is biased so that a pulse applied to either R1 or R2 will drive the device into conduction and produce an output pulse.

 With reference to Fig. 12–7, an AND gate is somewhat similar to an OR gate, except that the transistor is biased into collector saturation. In turn, an input pulse applied to R1 is insufficient to drive the transistor out of saturation. On the other hand, if an input pulse is simultaneously applied to R2, the sum of the two inputs will drive the transistor out of saturation and into cutoff. Therefore, an output pulse is produced, and this pulse has the same polarity as the input pulse, inasmuch as a CB configuration is employed. Next, when a CE configuration is utilized, as depicted in Fig. 12–7(b), the same general

(a)

(b)

Figure 12–8. Example of a pulse generator that produces clock pulses. **(a)** Basic circuit; **(b)** Collector output waveforms. *(Courtesy of Libe Company)*

circuit action is obtained, except that the output pulse is reversed in polarity with respect to the input pulse. In turn, this is called a NAND (NOT AND) gate.

Another basic "building block" is the pulse generator, or clock, which is the primary source of synchronizing signals in electronic computers. A typical pulse generator is shown in Fig. 12–8. It is essentially a square-wave generator that provides 6-volt output excursions. This is an example of a low-frequency (low repetition rate) generator that produces approximately two pulses per second. We shall find pulse generators in elaborate digital computers that operate at very high frequencies. However, the circuit-action principles are the same in all computers. Note in passing that Figs. 12–6 and 12–7 are examples of resistor-transistor logic, also called RTL logic. Other equivalent arrangements employ diodes and transistors and are called DTL logic. A widely used gate design utilizes transistors only, and is called TTL logic. A variation of TTL that provides extremely high speed of operation uses emitter-coupled transistors and is called ECL logic.

12 . 2 BASIC DIGITAL LOGIC

Logic symbols are to digital integrated circuits as schematic diagrams are to resistors, capacitors, and so on. Some of the more common types of logic symbols are depicted in Fig. 12–9. Most computers work on the binary principle. Binary stands for "two"; digital logic has two states or conditions, which are either ON or OFF (HIGH or LOW). Consider the condition in which we have zero or near zero volts at the input to a gate, flip-flop, amplifier, etc. This is an OFF condition. Next, if this voltage goes positive, let us say to 1 or 2 volts, it is now in an ON condition. There are two types of logic— positive logic and negative logic. A distinction is made as shown in Fig.

OR gate AND gate Amplifier Flip-flop

Figure 12–9. Common symbols for digital logic devices.

Figure 12–10. Distinction between positive logic and negative logic.

12–10 between the two kinds of logic. Thus "A" is denoted by "ON" or a binary "1" in positive logic, and NOT A ("A") is denoted by "OFF" or a binary "0" in positive logic. On the other hand, "A" is denoted by "OFF" or a binary "0" in negative logic, and NOT A ("A") is represented by "ON" or a binary "1" in negative logic.

Widely used functions in digital integrated circuits are as follows:

GATES. Control the passage of signals.

BUFFERS (AMPLIFIERS) (DRIVERS). Amplify the power of a signal either for driving heavy loads, or for sectional isolation.

INVERTERS. Reverse the logic state from positive to negative, or from negative to positive.

EXPANDERS. Permit additional inputs to a gate. For example, an expander allows a 3-input gate to be converted into a 7-input gate.

ADDERS. Provide SUM and CARRY (addition) operation on a pair of input signals.

SHIFT REGISTERS. Provide bistable storage of digital signals.

FLIP-FLOPS. Provide counting or division by 2, 4, 8, and so on.

Gate functions are encountered in a wide variety of configurations. Thus, two, three, four, or more inputs may be provided in four categories, as follows:

AND: When all inputs go to "1," the output will go to "1."

NAND: The output will be "1" except when all inputs go to "1."

OR: When any input goes to "1," the output will be "1."

NOR: The output will be "1" except when any input goes to "1."

Figure 12–11. Basic gate input/output relations.

Otherwise stated, NAND and NOR differ from AND and OR in that inversion has taken place. Refer to Fig. 12–11. Notice that a small o appears at the input or output of some of the examples. This o indicates that inversion takes place. Gates can be connected to operate in a wide variety of applications other than those for which they were basically designed. As an illustration, gates can be interconnected to operate as free-running (astable) multivibrators, bistable (flip-flop) multivibrators, and one-shot multivibrators.

An amplifier is referred to in digital work as a *buffer*. Its original use was to increase "fan-in" or "fan-out" capability. In other words, the number of other units that can be connected in parallel to the input (fan-in) or to the output (fan-out) is increased by the use of a buffer. Next, it should be remarked that there are many forms of flip-flops with which the experienced digital troubleshooter must become familiar. For example, the reset-set (R-S) flip-flop will be encountered. A slightly more elaborate type of flip-flop is the J-K arrangement, depicted in Fig. 12–12. A J-K flip-flop has a clock input, identified by T on the logic symbol. A comprehensive understanding of flip-flop action requires considerable study, and the reader

Gates connected as a
bistable flip-flop

Typical J-K flip-flop

Figure 12–12. Flip-flop consisting of gates, and the J-K logic symbol.

is referred to specialized digital theory textbooks such as *Digital Computers* by Deem, Muchow, and Zeppa, published by the Reston Publishing Co., Inc.

12.3 TROUBLESHOOTING DIGITAL CIRCUITRY

Troubleshooting digital circuitry is fundamentally based on testing for normal responses from gates and flip-flops, and checking for normal clock operation. In other words, suppose that an OR gate falls under suspicion. A basic test approach is to apply an input pulse to the gate, and to observe whether an output pulse is produced. Special test equipment is available for making this type of test. As noted previously, most digital circuitry employs IC's; each IC usually contains numerous gates, and may also contain flip-flops. A digital technician must understand the contents of a digital IC before he can apply test equipment intelligently. Accordingly, the following procedural discussion assumes that the troubleshooter has consulted the digital service data, or has looked up the particular IC in a digital IC manual. He knows, in other words, the gates and/or flip-flops that are contained in the IC, and the terminal connections to each.

A logic probe, illustrated in Fig. 12–13, is the most important single test instrument used in digital-circuitry troubleshooting. It is employed to trace logic levels and pulses through integrated circuitry to determine whether the point under test is logic high, logic low, bad level, open-circuited, or pulsing. This probe has preset logic thresholds of 2.0 and 0.8 volts, which correspond to the high and low states of TTL and DTL circuitry. When the probe is touched to a high-level point, a bright band of light appears around the probe

Figure 12–13. Logic probe used to troubleshoot digital circuitry. *(Courtesy of Hewlett-Packard)*

tip. When touched to a low-level point, the light goes out. Open circuits or voltages in the "bad level" region between the preset thresholds produce illumination at half brilliance. Single pulses of 10 μs or greater widths are made readily visible by stretching to one-twentieth second. The lamp flashes on or blinks off depending upon the polarity of the pulse. Pulse trains up to 50-MHz repetition rates cause the lamp to blink off and on at a 10-Hz rate.

Fig. 12–14 shows the logic probe in use. The circuit under test can first be operated at normal speed, while checking for the presence of key signals, such as clock, reset, start, shift, and transfer pulses. Next, the circuit under test can be stepped one pulse at a time, to determine whether a normal output pulse is produced. In this type of troubleshooting procedure, it is helpful to also employ a logic pulser, as illustrated in Fig. 12–15. This pulser provides a convenient means of stepping a circuit one pulse at a time, with the result monitored with the logic probe. Fig. 12–14 shows the pulser and logic probe in use. No adjustments of pulser or probe are required, and the trouble-

Figure 12–14. Logic probe in use. *(Courtesy of Hewlett-Packard)*

Figure 12–15. Logic pulser used to troubleshoot digital circuitry. *(Courtesy of Hewlett-Packard)*

shooter is freed to concentrate on circuit analysis instead of measurement techniques.

As a practical feature, connectors provided with the logic probe facilitate connection to the required 5-volt supply from either the circuit under test or from a separate supply. A ground clip is also provided which may be used to directly connect the probe circuitry to the ground of the circuit under test. Fig. 12–16 shows a block diagram of the logic-probe circuitry and its response to different inputs.

When using the logic pulser shown in Fig. 12–17, its tip is touched to the circuit under test, and the pulse button is pressed. In

Figure 12–16. Example of pulser and logic-probe application. *(Courtesy of Hewlett-Packard)*

turn, all circuits connected to the node (outputs as well as inputs) are briefly driven to their opposite state. No unsoldering of IC outputs is required. Since the pulse injection is automatic, the troubleshooter need not concern himself with whether the test node is in the high or low state. High nodes are pulsed low, and low nodes high, each time that the button is pressed. The pulser is essentially a single-shot pulse generator in probe form. An ability to source or sink up to 0.65 ampere ensures sufficient current to override IC outputs in either the high or low state. An output pulse width of 0.3 μs limits the amount of energy delivered to the device under test, thereby eliminating the possibility of damage.

Another very useful digital test instrument is the logic clip, illustrated in Fig. 12–18. This unit clips onto TTL or DTL IC's and instantly displays the logic states of all 14 or 16 pins. Each of the clip's light-emitting diodes independently follows level changes at its associated pin; a lighted diode corresponds to a high logic state. The clip contains its own gating logic for locating the ground and the +5-

volt V_{cc} pins. As a practical procedural note, a logic probe is much easier to use than an oscilloscope or voltmeter when the question is whether a lead is in the high or low (1 or 0) state. Pulser-probe and pulser-clip combinations enable the troubleshooter to quickly answer such questions as "Is a gate functioning?" or "Is a pin shorted to ground or V_{cc}?", and "Is a counter counting?" without unsoldering pins or cutting printed-circuit conductors.

Logic probe block diagram

Probe response to different inputs

Figure 12–17. Block diagram of the logic-probe circuitry and its response to different inputs. *(Courtesy of Hewlett-Packard)*

Figure 12–18. Logic clip used to troubleshoot digital circuitry. **(a)** Appearance of clip; **(b)** Logic clip in use. *(Courtesy of Hewlett-Packard)*

Gate operation is tested with the pulser driving the input and the probe monitoring the output pulses. When pulses do not appear at the output, the pulser and probe can be applied to the same pin to determine if the failure is due to a short to ground or V_{cc}. Testing sequential circuits such as flip-flops is accomplished with the clip simultaneously monitoring all output states while the pulser applies reset pulses to the device. Faulty operation becomes immediately apparent since the IC will not go through its prescribed sequence of states.

12 . 4 OSCILLOSCOPE APPLICATION

An oscilloscope is also essential in various areas of digital-circuitry troubleshooting. Figure 12–19 illustrates a 50-MHz general-purpose oscilloscope with triggered time base, which is suitable for trouble-shooting all types of digital circuitry. Erratic operation of gates and flip-flops can be caused by a seriously distorted pulse waveform, as

Figure 12–19. A 50-MHz oscilloscope suitable for digital-circuitry troubleshooting. (*Courtesy of Hewlett-Packard*)

exemplified in Fig. 12–20. Similarly, pulse repetition rates and pulse amplitudes can be accurately measured with a scope that has a calibrated time base. Triggered sweep is an essential feature. Note that high vertical sensitivity is not required, because pulse amplitudes are normally in the order of 2 volts peak-to-peak. A low-capacitance probe is necessary to provide high input impedance to the scope, and thereby avoid objectionable circuit loading. Direct current response is desirable, although the majority of tests can be made with an ac oscilloscope.

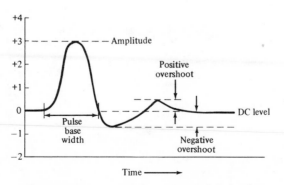

Figure 12–20. A seriously distorted digital pulse.

12.5 LABORATORY PULSE GENERATORS

Laboratory pulse generators are often useful in digital-circuitry trou-
bleshooting procedures. Such pulse generators are basically similar to
square-wave generators, except that a rectangular output waveform
is provided. In other words, the duty cycle (ratio of "on" and "off"
times) of a pulse generator is low, and may be very small in some
cases. The generator illustrated in Fig. 12–21 is called a two-channel
binary waveform generator, or simply a word generator. It produces
variable-length serial digital words at rates up to 10 MHz. Selections
of two 16-bit word sequences are available, and a single action com-
bines these two 16-bit words in series to provide a 32-bit word at
the output of both channels. A remote programming feature permits
rapid conversion of parallel words to serial words. Figure 12–22 de-
picts typical output pulse waveforms.

12.6 ANALYSIS BY SUPPLY-VOLTAGE
VARIATION

Supply-voltage variation provides a convenient troubleshooting tech-
nique whereby the margin of safety is determined between normal
operating conditions and the failure level. Power supplies of most
digital equipment are electronically regulated and adjustable over a
±10 percent range. In any case, the troubleshooter simply varies the
power-supply voltage to the area of concern both above and below
the rated value. The equipment waveforms are then checked with an
oscilloscope to determine whether operation is thereby disturbed. As
an illustration, if an inverter transistor has a deteriorated beta value,
its output waveform will stop when the power-supply voltage is in-

Figure 12–21. A pulse generator suitable for digital-computer circuitry tests.
(Courtesy of Hewlett-Packard)

Figure 12–22. Typical pulse waveforms used in digital-computer troubleshooting. *(Courtesy of Hewlett-Packard)*

creased past the critical value, because the base drive will then be inadequate to provide normal operation. Again, reduction of power-supply voltage reveals the possible effects of noise on system response, or detects a condition wherein the input-signal swing is inadequate.

12 . 7 CIRCUIT-BOARD TESTERS

Unlike faults in linear circuitry, most defects in digital circuitry consist of deteriorated or nonoperative devices, broken wires or conductors, and short-circuits. When a system has more than one fault, direct substitution of a new circuit board is not necessarily the best approach. If a circuit-board tester is available, comparison tests of suspected boards can be quickly made against new or known-good boards. As exemplified in Fig. 12–19, a circuit-board tester comprises a pair of circuit-board sockets for insertion of the suspected board and the reference board, together with switching facilities and an ohmmeter. Note that practically any circuit board can be given a comparative test with this arrangement. Note that various manufacturers sometimes employ different types of circuit-board sockets, and the tester should be constructed accordingly.

 With reference to Fig. 12–23, when no pushbutton switch is pressed, all terminals of both sockets are connected together. On the other hand, if any one button is pressed, the socket terminal associated with that button is disconnected from the common bus and connected to the other bus line. Note that the ohmmeter is connected via polarity-reversing toggle switch S1 between these two bus lines. Thereby, the resistance at any terminal of the suspected circuit board can be compared with the resistance at the corresponding terminal of the reference circuit board. This resistance comparison can be made

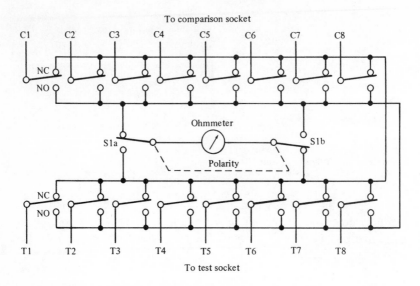

Figure 12–23. Typical circuit-board tester (for boards with eight terminals).

with both polarities of ohmmeter voltage. Note that the technician is not concerned with the internal circuitry of the board, because this is basically a go/no-go type of testing.

REVIEW QUESTIONS

1. Describe modular construction of digital circuitry.
2. What are the three chief "building blocks" in an electronic computer?
3. Explain the function of a steering diode.
4. Define a digital memory; a digital counter.
5. Distinguish between an AND gate and an OR gate; between a NAND gate and a NOR gate.
6. Discuss the function of a buffer; an inverter; an expander; an adder.
7. State the principle on which digital troubleshooting is fundamentally based.
8. Explain the function of a logic probe; how is it applied in digital circuitry?

9. Describe the function of a logic pulser; how is it utilized in troubleshooting procedures?

10. Discuss the function of a logic clip; what troubleshooting questions does a logic clip answer?

11. What kind of oscilloscope is required for troubleshooting digital circuitry?

12. Explain the utility of a low-capacitance probe.

13. Would you use a logic probe or an oscilloscope to check pulse repetition rates and pulse amplitudes?

14. Could you use a logic probe to check pulse distortion?

15. Why is triggered sweep an essential feature of a digital troubleshooting oscilloscope?

16. Is high vertical sensitivity required in an oscilloscope used for digital troubleshooting?

17. Is it necessary to use an oscilloscope with dc response to check digital circuits?

18. Explain the relation between a pulse generator and a square-wave generator.

19. Why is a pulse generator useful in digital troubleshooting procedures?

20. Discuss the supply-voltage variation technique of digital troubleshooting.

21. Are the power supplies in most digital equipment adjustable?

22. Name the most common faults encountered in troubleshooting digital systems.

23. Is substitution of a suspected circuit board always the best approach to digital trouble symptoms?

24. Explain how circuit-board resistance comparison tests may be made.

Answers to
Review Questions

CHAPTER 1

1. A thermistor functions to sense changes in temperature in terms of resistance variation.
2. A heat sink is applied between the point of soldering and the case of the device.
3. An anti-wicking tool prevents solder from flowing under the insulation of a wire during soldering operations.
4. Electrician's all-purpose pliers can be used to cut screws, to cut wire, to strip wire, or to crimp lugs.
5. The primary advantage of modular construction is that troubleshooting procedures are speeded up by substitution tests with new modules.
6. A coating of silicone grease is applied between a power transistor and its heat sink for optimum heat transfer.
7. See Figure 1–26.
8. When soldering semiconductors on PC boards, care must be taken to avoid overheating the devices and to avoid loosening pads from the PC board.
9. The advantage of clipping the leads of a defective component on a PC board and soldering a new component to the old leads is avoidance of possible damage to the PC board.
10. Use of a small spiral connector in the operation of question 7 adds mechanical strength to the connection.

11. Bipolar transistors have low input resistance compared to JFET's; a bipolar transistor is essentially current-operated, whereas a JFET is essentially voltage-operated.

12. A MOSFET has an insulated-gate electrode, whereas a JFET has a junction-gate electrode.

13. A MOSFET can be protected from damage by static electricity if its gate terminal is connected to its source terminal.

14. An SCR has a switching action that is essentially the same as that of a thyratron.

15. Out-of-circuit semiconductor testers check transistors for beta value and leakage; in-circuit testers can show only whether a transistor is workable.

16. Semiconductor curve tracers automatically plot voltage-current characteristics of a device on a scope screen.

17. Collector-junction leakage in a transistor is equivalent to a resistive path between the collector and base regions of the device.

18. Solder that has been collected by a desoldering tool becomes oxidized and will not make a normal bond.

19. "Modular construction" denotes circuit boards that can be plugged and unplugged for quick checks of equipment operation.

20. Heat sinks utilized with power transistors provide rapid dissipation of heat so that device damage is avoided.

21. Diode polarity may be indicated by a + sign, by a black band, or by a colored dot.

22. A typical integrated circuit has 12 terminals.

23. Excess solder should be removed from a PC board by means of a desoldering tool.

24. If a hole in a PC board becomes filled with solder, the solder should be melted and pushed out with a length of wire.

CHAPTER 2

1. The term "dc voltage distribution" denotes circuit-voltage values during normal operation of a unit.

2. An oscilloscope displays the waveshape of an ac voltage, in addition to its peak-to-peak voltage.

3. A CE circuit has its supply and bias voltages returned to the emitter of the transistor; return is to the base in a CB circuit; return is to the collector in a CC circuit.

4. Any point in a circuit that is grounded is a common-ground point for all the branch circuits to which it connects.

5. A CE circuit may have an arbitrary common-ground point.

6. Germanium transistors have a typical forward bias of 0.3 volt; silicon, 0.6 volt.

7. Transistor defects usually result in dc-voltage distribution changes.

8. An open collector junction causes the collector terminal to rise to the supply-voltage value; an open emitter junction causes the base terminal to rise to the collector potential.

9. Typical symptoms of a leaky collector junction are reduced collector voltage with increased base and emitter voltages.

10. If an emitter junction is short-circuited, the base and emitter terminals rest at the same potential and collector current is cut off.

11. A "floating" electrode has its terminal open-circuited.

12. Normal ranges of transistor operating voltages for a small-signal germanium transistor are V_{BE} from 0.05 to 0.3 volt; V_{CE} from 2 to 9 volts.

13. Normal ranges of transistor operating voltages for a small-signal silicon transistor are V_{BE} from 0.6 to 5 volts; V_{CE} from 10 to 60 volts.

14. Class-A operation of a transistor denotes linear input-output relations; in class-B operation, the transistor is cut off for one-half of the operating cycle; in class-C operation, the transistor is cut off for more than one-half of the operating cycle.

15. A sine wave is the most fundamental waveform.

16. In a sine wave, the rms value is equal to 0.707 of peak, and the peak-to-peak value is equal to double the peak value.

17. "Cutoff" in a transistor denotes zero collector current; "saturation" denotes heavy emitter-current flow with very low collector voltage—the collector junction becomes forward-biased at saturation.

18. Waveform distortion is measured in terms of percentage harmonic distortion.

19. The chief types of square-wave distortion are rounded corners, slow rise, curved top and bottom excursions, tilt, overshoot, and ringing.

20. Delay time between input and output signals denotes that the output waveform lags the input waveform.

21. Storage time in a transistor is a measure of the time required for the stored carriers in the base region to be swept out after the device has been in saturation.

22. A frequency marker is produced on a frequency-response curve by mixing a CW signal with the FM sweep signal.

23. Lissajous patterns are developed by applying one sine-wave voltage to the vertical channel of a scope, and applying another (integrally related) sine-wave voltage to the horizontal channel of the scope.

24. Lissajous patterns can show overload distortion, phase shift, nonlinearity, and crossover distortion.

CHAPTER 3

1. Complementary push-pull action in an amplifier denotes alternate amplification by PNP and NPN transistors, each of which develops a half cycle of the output waveform. These half cycles are combined in the load to provide the full cycle of operation.

2. The CE mode of operation provides the most nearly equal input and output resistances.

3. Voltage gain is defined as the ratio of base signal voltage to collector signal voltage; current gain is defined as the ratio of base signal current to collector signal current; power gain is defined as the product of voltage gain and current gain.

4. A high-frequency rolloff network provides treble cut.

5. A low-frequency boost network provides bass boost.

6. Degenerative feedback stabilizes transistor operation and improves operating linearity at the expense of stage gain.

7. A disadvantage of circuit loading by a voltmeter is inaccurate scale indication.

8. When electrolytic capacitors are utilized, they are the most likely cause of amplifier malfunction.

9. In a turn-off test, the base and emitter are brought to the same potential and the collector voltage is measured. In a turn-on test, added forward bias is bled into the base from the collector of the transistor, and the collector voltage is measured.

10. Circuit arrangements may be such that it is impractical or impossible to make a turn-off test of a transistor.

11. A lo-pwr ohmmeter should be used, and sometimes must be used

to make in-circuit measurements of resistor values in a solid-state amplifier.

12. It is not possible to measure the value of a resistor in-circuit if it is shunted by a low value of resistance such as a coil winding.

13. A depletion-type FET operates with substantial current flow at zero bias; an enhancement-type FET has very small current flow at zero bias.

14. A suspected open-circuited capacitor can be quick-checked in-circuit by bridging with a known good capacitor.

15. To avoid possible damage from current surges, the equipment should be turned off before a suspected capacitor is bridged.

16. Standard resistor color-code values are: black, 0; brown, 1; red, 2; orange, 3; yellow, 4; green, 5; blue, 6; violet, 7; gray, 8; white, 9.

17. Capacitor color codes are not completely standardized.

18. A power resistor is substituted for the speaker in amplifier servicing procedures to avoid the annoyance of loud sound reproduction.

19. To signal-trace an audio amplifier, a 1-kHz signal voltage is usually applied, and the sine waveform is checked at progressive stages with an oscilloscope.

20. A technician cannot "see" 1% harmonic distortion in a sine wave displayed on the screen of an oscilloscope.

21. The most common cause of harmonic distortion in an audio amplifier is incorrect biasing of one or more transistors.

22. A negative-feedback loop in an amplifier feeds back a fraction of the output waveform to the input, where it combines in opposing phase with the input waveform.

23. Negative-feedback loops make evaluation of trouble areas comparatively difficult in signal-tracing procedures.

24. When troubleshooting intermittents, a technician generally monitors the voltages or waveforms at various points in the equipment, and endeavors to "speed up" the occurrence of the intermittent by application of heat or cold or tapping of suspected components or devices.

CHAPTER 4

1. "Ripple" denotes the AC component that is present in the DC output from a power supply.

2. A 10% ripple is regarded as a trouble symptom in most types of electronic equipment.

3. A half-wave power supply develops a 60-Hz ripple voltage.

4. A full-wave power supply develops a 120-Hz ripple voltage.

5. Successive peaks in a ripple waveform will have unequal amplitudes if one of the rectifier diodes in a full-wave arrangement has a poor front-to-back ratio.

6. An open-circuited diode in a full-wave power supply causes the circuit to operate as a half-wave supply.

7. "Regulation" is defined as the difference between the no-load output voltage and the output voltage under load, divided by the output voltage under load.

8. The average value of the output voltage from a half-wave rectifier is 0.318 of peak value.

9. A full-wave rectifier provides an average output-voltage value of 0.636 of peak value.

10. Ripple voltage can be measured approximately on the Output function of a TVM.

11. A full-wave bridge-rectifier configuration employs four rectifiers that are connected to provide a full-wave output from a secondary winding that is not center-tapped.

12. When a power supply is switched on, a surge-current demand is placed on the rectifier diode to charge the input filter capacitor.

13. Analysis of an overcurrent trouble symptom in a power supply is made by operation at reduced input voltage, or by disconnecting the load branches and reconnecting the branches one-by-one.

14. A dual-polarity power supply employs two rectifier diodes in opposite polarities, and two separate filter circuits.

15. A loud buzzing sound from a power transformer indicates a serious overload condition.

16. "Peak inverse voltage" denotes the double value of output voltage that the rectifier diode must withstand during the non-conducting half-cycle.

17. Circuit breakers are classified into thermal types and magnetic types.

18. A voltage-multiplier power supply employs successive voltage-double circuits, with the result that increased output voltage is obtained at the expense of regulation.

19. Electrolytic capacitors are most likely to cause malfunction in a power supply.

20. A regulated power supply maintains a comparatively constant output voltage under conditions of varying output current demands.

21. Zener diodes are connected in opposition to the load when a regulated output of less than 3.9 volts is required.

22. The two general classes of Zener-diode/transistor regulated power supplies are termed series-regulated and shunt-regulated configurations.

23. Amplified control action in a regulated power supply entails step-up of the sensed error voltage prior to its application in the regulating circuit.

24. The Darlington transistor configuration employs an input transistor in the CC configuration that is directly coupled to an output transistor which also operates in the CC configuration.

CHAPTER 5

1. An SCR is a DC type of controlled rectifier; a diac is a two-lead AC type of controlled rectifier; a triac is a full-wave AC type of controlled rectifier.

2. A basic motor speed-control unit functions by current variation which is determined by the state of an associated SCR, diac, or triac.

3. The function of a snubber circuit is to reduce the peak voltage of transients.

4. Semiconductor devices are most likely to become defective in a motor speed control unit.

5. A conventional diac and triac can control the speed of an ac motor rated for 8-ampere operation.

6. Electronic timers typically employ an RC time-constant circuit that switches a triac.

7. Incorrect timing action in an electronic timer can be caused by a worn potentiometer or a drift in fixed-resistor values; no response throws preliminary suspicion upon the power supply; erratic operation may result from loose connections or a worn timing control; intermittent operation can result from the foregoing faults or from thermally responsive capacitors or transistors.

8. It would be practical to set an MOS electronic timer for a period of 11 seconds.

9. A single triac may be employed in a lamp-dimmer control unit.

10. Abrupt conduction of a triac can cause RF interference.

11. RF interference can be confined to a triac unit by means of an LC line filter.

12. Neon lamps eventually become inoperative owing to change in gas pressure.

13. Triacs are utilized in off-on heat controls with synchronous switching.

14. A two-transistor regenerative switch employs a flip-flop configuration wherein positive feedback provides very rapid switching action.

15. If substantial voltage is measured at the gate of a triac when the device is supposed to be off, it is reasonable to conclude that the triac is defective.

16. A thermistor has a negative temperature coefficient of resistance; its resistance value decreases rapidly with an increase in temperature.

17. Positive action in a relay configuration denotes that the load will receive full power or no power, depending upon sensor threshold settings.

18. A photocell converts light energy into electrical energy.

19. If a control unit remains in one state regardless of input actuating voltage, it is said to have a "stuck-at" trouble symptom.

20. A photocell has a typical resistance variation from 100 ohms to many megohms as the light input varies from an intense level to total darkness.

21. A diac has two terminals which are not separately identified; a triac has three terminals identified as main terminal 1, main terminal 2, and the gate.

22. Switching operations are provided by a unijunction transistor in response to trigger pulses; amplification is also provided, and the device can be used as an oscillator.

23. Selenium and silicon are utilized in photovoltaic cells.

24. A silicon photocell can develop an output of 0.5 volt at 40 mA.

CHAPTER 6

1. Transistor voltmeters are comparable to volt-ohm-milliammeters with a bridge-type amplifier that provides high input impedance and high sensitivity.

2. Calibration is the process of checking an instrument against a standard to adjust the instrument response for optimum accuracy.

3. If a TVM is off-calibration over part of its dc-voltage ranges, at least one of its multiplier resistors is probably off-value.

4. When the ohmmeter function of a TVM becomes off-calibration, the ohmmeter battery is suspected first.

5. In case a TVM is inoperative except for its dc-current function, transistor-bridge operation would be checked.

6. If an ohmmeter is accidentally connected across a "live" circuit, the associated multiplier resistors are likely to be damaged; the input transistor may also be damaged.

7. The chief sections of a service-type oscilloscope are the CRT, vertical amplifier, horizontal amplifier, time base, and power supplies.

8. Typical service-type oscilloscopes have a vertical-amplifier bandwidth of 5 MHz and a voltage gain of 1500 times.

9. From 1000 to 2000 volts are commonly applied to the CRT in a service-type oscilloscope.

10. Typical horizontal amplifiers in service-type oscilloscopes have a bandwidth of 1 MHz and a voltage gain of 50 times.

11. If a 0.71V sine wave is applied to the vertical-input terminal of a scope, and the vertical-gain control is adjusted for a screen deflection of 2 inches, the scope is then calibrated for a sensitivity of 1 peak-to-peak volt per inch.

12. The most probable cause for waveform clipping in the vertical amplifier of an oscilloscope is incorrect bias on an amplifier transistor.

13. Incorrect bias on a horizontal-amplifier transistor can cause nonlinear horizontal deflection.

14. Positioning controls of an oscilloscope can drift out of normal range owing to leakage in coupling capacitors.

15. Poor focus is generally caused by disproportionate voltages on the CRT anodes.

16. If the beam intensity becomes weak in a CRT, the accelerating voltage may be subnormal, or the cathode may be nearing the end of its useful life.

17. Oscillation in the vertical amplifier of an oscilloscope is ordinarily caused by "open" decoupling capacitors.

18. Loss of sync action in an oscilloscope throws suspicion upon coupling capacitors in the sync channel.

19. A substitution test is usually made when a CRT falls under suspicion.

20. A saturated transistor on one side of a differential amplifier will sometimes appear to be a trouble condition on the other side of the amplifier.

21. A bypass or decoupling capacitor that is suspected of being "open" can be quick-checked by bridging with a known good capacitor.

22. Transistor turn-off tests can usually be made in dc-coupled circuitry, although it may be found advisable to check the transistors by pairs.

23. When astigmatism occurs, the focus control cannot be adjusted to obtain a properly focused display at both center-screen and edge-screen.

24. When measuring the high-voltage power-supply potential in an oscilloscope, the technician should remember that the heater and cathode of the CRT are probably 1000 to 2000 volts above ground, and that the accelerating anode is at ground potential.

CHAPTER 7

1. Typical stage gains for a conventional AM superheterodyne receiver are: RF amplifier, 10X; converter, 25X; IF amplifier, 85X; audio driver, 80X; audio output, 15X.

2. Visual inspection of a receiver might reveal a cracked circuit board, loose leads, a cracked loopstick antenna, excessive moisture, or breaks at the tuning-capacitor terminals.

3. Accidental reversal of the supply-battery polarity is likely to burn out transistors and damage electrolytic capacitors.

4. A current-drain measurement is made by connecting a milliammeter in series with a lead to the supply battery.

5. A typical AM radio receiver has a converter current drain of 0.5 to 1mA; each IF stage, 1.5 mA; audio driver, 1.5 mA; audio output stage, 4 mA.

6. If the current drain in a battery radio is zero, the off-on switch falls under preliminary suspicion.

7. Excessive current drain may be caused by a faulty transistor or by a short-circuited filter capacitor.

8. A "click" test is made in a radio receiver by flipping the off-on switch while listening for a clicking sound from the speaker.

9. To quick-check the local oscillator in an AM radio receiver place it near another receiver tuned to a station at the high-frequency end of the band. Then, tune the "dead" receiver through its range; if the oscillator is operating in the "dead" receiver, a heterodyne whistle will be heard.

10. AGC action can be checked by turning the volume control to minimum and measuring the battery current while the receiver is tuned through its range. If the AGC section is operating, a dip will be noted in the meter reading when the receiver is tuned to a station.

11. Signal-injection tests include application of a 1 kHz sine-wave signal at the speaker, at the input of the audio-output stage, and at the output of the audio-driver stage; application of a modulated-IF signal at the input of the detector, and at the input(s) of the IF stage(s); and application of a modulated-RF signal at the input of the converter stage and at the input of the RF stage (if used).

12. It is impractical to measure the base current of a transistor with a VOM because the internal resistance of the instruments disturbs the bias voltage.

13. The output level from a signal generator can be checked with a field-strength meter.

14. A generator signal is inductively coupled to a loopstick antenna by means of a coil of wire with a diameter of approximately 2 inches and about a dozen turns.

15. A signal tracer is used with a detector probe to check for signals in the RF and IF sections of a receiver; it is used with an audio probe to check for signals in the audio section of the receiver.

16. Signal-tracing procedures generally start at the converter, and proceed step-by-step toward the speaker in a radio receiver.

17. An AGC line is clamped with battery voltage, or with the output from a bias supply box. The receiver service data usually specify the normal AGC voltage.

18. A weak-output symptom can be analyzed to best advantage by making stage-gain measurements. This is a quantitative aspect of signal-injection tests.

19. Typical distortion trouble symptoms include mushy or raspy sound, whistles, motorboating, squeals, and tweets.

20. Preliminary distortion localization can often be made by observing the change or lack of change in the distortion as the volume con-

trol is varied, and observing any change in distortion from weak-signal stations to strong-signal stations.

21. Common causes of audio distortion are incorrect transistor bias, defective transistors, incorrect detector bias, or "open" decoupling capacitors.

22. Regeneration trouble symptoms include squeal, hiss, motorboating, and unusually critical tuning.

23. Poor ground connections are often responsible for regeneration in a radio receiver.

24. The four-step peak-alignment procedure for a radio receiver consists of peaking each tuned-IF circuit at 455 kHz, adjusting the oscillator tank for optimum reception at 1500 kHz, adjusting the oscillator padding capacitor for optimum reception at 600 kHz, and adjusting each tuned-RF circuit for maximum response at the frequency specified in the receiver service data.

CHAPTER 8

1. Two general sources of trouble symptoms in FM receivers are component defects and drift in alignment adjustments.

2. Basic sections of a combination receiver include an FM tuner, an AM tuner, an IF section, stereo-multiplex decoder, and audio section.

3. Preliminary trouble analysis in a combination receiver entails checking response on each function, such as FM mono, FM stereo, AM, and phono or tape operation, if provided.

4. A ratio detector is often checked out with a signal generator and a TVM for center-frequency and limit-frequency response.

5. If an oscilloscope has 10-MHz bandwidth, it can be used to check the IF amplifier, stereo-multiplex decoder, and audio sections of an FM receiver.

6. A demodulator probe operates as a traveling detector; it demodulates a high-frequency signal for processing by a scope with comparatively limited frequency response.

7. "Incidental frequency modulation" denotes the spurious FM component that is produced when the oscillator in an AM signal generator is directly modulated.

8. The basic sections utilized in a stereophonic FM receiver include the tuner, IF amplifier, decoder, audio amplifier, and twin speakers.

9. Three basic stereo-multiplex decoding arrangements are the envelope-detector method, bandpass-and-matrix method, and the electronic-switching method.

10. The most basic test of decoder operation is a measurement of L and R channel separation in dB.

11. Separation is checked with the aid of a stereo-multiplex generator and a TVM or oscilloscope. It is customary to express separation in decibels.

12. A separation of 10 dB is regarded as a trouble symptom in a stereophonic FM receiver.

13. An L+R signal differs from an L−R signal by 180° in phase.

14. A pilot subcarrier has a frequency of 19 kHz, and its phase is precisely related to the 38-kHz subcarrier that is suppressed at the transmitter.

15. Subcarrier suppression is employed to improve the signal-to-noise ratio of the transmission.

16. Preemphasis is utilized to optimize the signal-to-noise ratio of the transmission.

17. If deemphasis were used prior to decoding at the receiver, the higher frequencies of the stereo signal would be seriously attenuated, and the decoder would fail to operate normally.

18. The audio-frequency range employed in stereo-FM broadcasting is from 20 Hz to 20 kHz.

19. The bandwidth utilized in an FM-IF amplifier is approximately 0.2 MHz.

20. An RF frequency range from 88 to 108 MHz is used in stereo-FM broadcasting.

21. An L+R signal is the equivalent of an monophonic FM signal because it consists of a simple mixture of the sounds picked up by the L and R microphones.

22. A center frequency is 10.7 MHz is employed in an FM-IF amplifier.

23. An IF stage in a combination receiver can process either FM or AM signals because the 10.7-MHz transformer windings "look like" short-circuits at 455 kHz, and the trimmer capacitors of the 455-kHz transformers have very low reactance at 10.7 MHz.

24. A blocking capacitor is used in series with the "hot" lead of a generator in signal-tracing tests to prevent drain-off of bias voltage at signal-injection points in the receiver under test.

CHAPTER 9

1. "Quadriphonic" denotes a system whereby sound picked up by four separate microphones is recorded on separate channels and played back through separate channels driving their own speakers.

2. Discrete quadriphonic operation employs four channels throughout the system; synthesized quadriphonic operation utilizes two channels in the initial stage and develops four channels in the final stage.

3. Two sources of quadriphonic sound are tapes and disc recordings.

4. A decoder is required for reproduction of quadriphonic records because the front and rear channels are encoded into the left and right channels.

5. Quadriphonic synthesizers develop front and rear sound signals by picking out certain left and right signal components on the basis of frequency and phase.

6. A gain-riding circuit automatically enhances front and rear separation when small separations are sensed.

7. FM broadcast stations do not transmit discrete quadriphonic sound because of bandwidth limitations imposed by the FCC.

8. The established FM broadcast channel bandwidth is 0.2 MHz.

9. A 30-kHz carrier frequency is utilized in quadriphonic discrete-type recordings.

10. When 180° phase-shift encoding is employed, more or less sound cancellation occurs at certain frequencies in particular positions of the listening area.

11. A disadvantage of the 90° phase-shift encoding technique is limited separation among channels.

12. An SQ system differs from a QS system chiefly in that the former decodes disc-recorded signals only.

13. QS systems are not entirely compatible with conventional stereo equipment.

14. Synthetic diffused-stereo reproduction occurs when a conventional stereo recording is played through a QS unit; opinions concerning this sound reproduction are equivocal.

15. A quadriphonic system may employ hi-fi L and R speakers, with utility-type F and R (rear) speakers.

16. An SQ quadriphonic decoder employs RC phase-shifting networks for decoding; digital-logic networks are utilized for gain-riding; matrixing action develops four separate signals that drive the speakers.

17. Optimum speaker placement must be experimentally determined in a quadriphonic installation because room acoustics and listener preferences vary considerably.

18. It is not always desirable to place speakers in the corners of a room.

19. A "live" room provides considerable sound reflection, whereas a "dead" room absorbs considerable sound energy.

20. The most basic quadriphonic performance test entails separation checks.

21. Separation is checked in a quadriphonic system with a demonstration record.

22. Demonstration records provide L, R, F, and R (rear) signals sequentially for separation tests.

23. Basic comparison tests in decoder troubleshooting involve dc-voltage, resistance, and waveform cross-checks.

24. A quadriphonic decoder cannot operate at less than 0.5% harmonic distortion in the present state of the art.

CHAPTER 10

1. Some of the solid-state devices used in industrial-electronics equipment are diodes, transistors, thermistors, photocells, varactors, varistors, silicon controlled rectifiers, diacs, triacs, and operational amplifiers.

2. The chief characteristics of an op amp are its high gain, freedom from offset voltages and currents, and amenability to operation with a variety of negative-feedback networks.

3. Op amps are applied in precision instrumentation amplifiers, signal conditioners, signal processors, signal generators, controlled-current sourdes, active filters, and comparators in industrial-electronic equipment.

4. An inverting output produces an output that is 90° out of phase; a non-inverting input produces an in-phase output.

5. An OVA has single-ended output, whereas an OTA has double-ended output and can drive a load that floats above ground.

6. A differential amplifier is characterized by two similar input circuits that respond to the difference between two input voltages or currents.

7. A constant-current source provides an output current that maintains virtually the same value, regardless of the load resistance.

8. A transistor with a fixed bias operates as a constant-current source because its collector current is determined primarily by the base-emitter bias voltage, and not by the collector voltage.

9. If the voltage-source polarity to an op-amp is accidentally reversed, the device is likely to be seriously damaged.

10. Series protective diodes are utilized to protect an op amp against supply-voltage polarity reversal.

11. A gain of 1000 times is not unusual for an op amp.

12. A virtual ground is established when a very large amount of negative feedback is employed with an op-amp; this feedback maintains the input terminal of the device virtually at zero signal potential.

13. System gain is calculated for an op amp that employs substantial negative feedback as the ratio of feedback resistance to input resistance.

14. Precise differentiating action is provided by an op-amp differentiator because the input capacitor "sees" a virtual ground.

15. Circuit actions that result in precise integrating action by an op-amp integrator are the constant-current charge/discharge characteristic owing to the virtual ground, and the extremely high gain of the device.

16. A basic op-amp integrating circuit is elaborated to prevent damage to the device by limiting the low-frequency characteristic of the network.

17. An op-amp summer is basically an analog adder that provides an output voltage equal to the sum of two or more input voltages.

18. An op-amp subtracter is basically an analog difference-calculator that provides an output voltage equal to the difference between two input voltages, or between a multiplicity of input voltages with various polarities.

19. "Scaling" or "weighting" entails automatic assignment of an arbitrary multiplying factor to a particular input signal for an op-amp summer.

20. An op-amp logarithmic amplifier employs nonlinear feedback resistance; an op-amp anti-logarithmic amplifier utilizes nonlinear input resistance.

21. Op-amps are configured in logarithmic/anti-logarithmic arrangements to perform multiplying and dividing operations.

22. In preliminary analysis, trouble symptoms may often be localized to the op amp or to the associated circuitry by unplugging the device and testing the associated circuitry.

23. Leakage between conductors of a PC board can simulate capacitor leakage, offset currents, or faulty op amps.

24. An output-offset voltage trouble symptom is approached by checking the setting of the output offset null adjustment.

CHAPTER 11

1. A television troubleshooter needs to be familiar with typical design features and variations so that he can make intelligent tests of circuit action.

2. A trouble symptom of no picture and no sound indicates a fault prior to video detection; no picture with sound normal indicates a faulty video-output stage or a defective picture tube; no sound with picture points to malfunction in the sound channel.

3. Loss of both picture and sound reproduction does not necessarily throw suspicion upon the video amplifier; the video detector could be at fault, for example.

4. A high snow level indicates trouble near the input of the signal channel; a moderate snow level corresponds to a fault near the input of the IF channel; a low snow level suggests a defect in the latter portion of the IF amplifier.

5. Noise-level outputs from the speaker correspond in a general way to snow-level trouble analyses.

6. A test-pattern generator facilitates preliminary symptom analysis by providing progressive injection signals with a meaningful screen display.

7. Three subsections of a VHF tuner are the RF amplifier, the oscillator, and the mixer.

8. A local oscillator is checked by dc-voltage measurements which are compared to the values specified in the receiver service data.

9. A class-C oscillator often appears to operate with a reverse-biased transistor because collector current flows in narrow pulses, with development of signal-rectified bias in the base-emitter circuit.

10. An IF signal can be "jumped" across an IF stage with a 500-pF capacitor connected from the output of the preceding stage to the input of the succeeding stage.

11. An AGC line is clamped with battery voltage, or with bias-box voltage, at the normal operating value specified in the receiver service data.

12. IF oscillation causes an abnormally high dc-voltage output to be developed by the picture detector.

13. IF oscillation can be caused by in many cases by peaking all of the IF tuned circuits to the same frequency.

14. If a replacement picture-detector diode is connected in reverse polarity, a negative picture is displayed, sync action is impaired, and AGC action is reversed.

15. Signal gain is checked in the video-amplifier section with an oscilloscope.

16. Video-amplifier malfunction is commonly caused by defective electrolytic capacitors; the video-output transistor is also a ready suspect.

17. A total AGC range of 70 dB is typical for a television receiver.

18. Vertical anti-hunt circuits function to stabilize the vertical oscillator-output system.

19. A dark screen points to lack of high voltage; a narrow picture points to low supply voltage; blooming corresponds to subnormal high voltage; horizontal nonlinearity is commonly caused by capacitor leakage; horizontal foldover is often the result of transistor leakage; keystoning generally points to a defective yoke.

20. Alignment is generally accomplished with the aid of a sweep and marker generator and an oscilloscope.

21. "Ground loops" result from excessively long RF ground circuits; instability results which can usually be corrected by using a very low-impedance ground plane such as a copper sheet, or by employing a common RF grounding point.

22. Overloading of the signal channel during alignment procedures results in display of distorted response curves and resulting misalignment.

23. Slugs in tuned-IF circuits should be positioned near the outer ends of the coupled windings—not near the inner ends of the windings.

24. Alignment of a TV receiver is usually the final procedure that is observed when troubleshooting any malfunction.

CHAPTER 12

1. Modular construction denotes the provision of circuit boards that can be plugged or unplugged without any attention to connecting leads.

2. The three chief building blocks of an electronic computer are the AND gate, the OR gate, and the astable or bistable multivibrator.

3. A steering diode directs an incoming trigger pulse to an appropriate transistor.

4. Digital memories comprise flip-flops (or equivalent arrangements) for storage of binary numbers. Digital counters indicate the sum of a train of incoming pulses in terms of binary numbers.

5. An AND gate provides a high output only if all of its inputs are high at the same time; an OR gate provides a high output if any one of its inputs is high; a NAND gate functions oppositely from an AND gate; a NOR gate functions oppositely from an OR gate.

6. Buffers provide amplification and isolation between various sections of a digital computer.

7. Digital troubleshooting is based fundamentally on testing for normal responses from gates and flip-flops, and checking the clock operation.

8. A logic probe indicates whether a terminal is in a high state or a low state; it is applied at any operative terminal in the digital-computer network.

9. A logic pulser is a single-shot pulse generator; it may be applied at any operative terminal in a digital-computer network, including a terminal to which a logic probe is simultaneously applied.

10. A logic clip indicates the states of all terminals of a digital IC simultaneously.

11. An oscilloscope used in digital troubleshooting procedures has a typical bandwidth of 50 MHz and a calibrated triggered time base.

12. A low-capacitance probe minimizes circuit loading by an oscilloscope input cable.

13. Pulse repetition rates and amplitudes are checked with an oscilloscope.

14. A logic probe cannot be used to check pulse distortion.

15. Triggered sweep is an essential feature of a digital troubleshooting oscilloscope to measure time intervals and pulse widths.

16. High vertical sensitivity is not required in an oscilloscope used for digital troubleshooting.

17. An oscilloscope used to check digital-circuit action does not require dc response.

18. A pulse generator is essentially a square-wave generator with a very low duty cycle.

19. A pulse generator is useful in digital troubleshooting procedures to apply single pulses, pulse trains, or digital "words" to systems under test.

20. Supply-voltage variation is often informative in digital-trouble-shooting procedures to uncover marginal operating conditions in various circuits.

21. Power supplies in most digital equipment systems are adjustable.

22. The most common faults encountered in troubleshooting digital systems are faulty devices and poor connections.

23. Substitution of a suspected circuit board is not always the best approach to digital trouble analysis.

24. Circuit-board resistance comparison tests are made with reference to another similar board that is in known good operating condition.

6.2

overhead an albatross caught a light of gold on its wings. Then Tavana Nui turned to his people and cried: "Here is my son come home from the sea. Mafatu, Stout Heart. A brave name for a brave boy!"

Mafatu swayed where he stood. "My father, I . . .

Tavana Nui caught his son as he fell.

It happened many years ago, before the traders and missionaries first came into the South Seas, while the Polynesians were still great in numbers and fierce of heart. But even today the people of Hikueru sing this story in their chants and tell it over the evening fires.

empty. Silence gripped them, and a chill of awe touched them like a cold hand. Then they saw a head lift above the gunwale, a thin body struggle to sit upright, clinging to the mid-thwart.

"*Auê te auê!*" The cry went up from the people in a vast sigh. So they might speak if the sea should give up its dead.

But the boy who dropped overside into the shallows and staggered up the beach was flesh and blood, albeit wasted and thin. They saw that a necklace of boar's teeth shone upon his chest; a splendid spear flashed in his hand. Tavana Nui, the Great Chief of Hikueru, went forward to greet the stranger. The brave young figure halted, drew itself upright.

"My father," Mafatu cried thickly, "I have come home."

The Great Chief's face was transformed with joy. This brave figure, so thin and straight, with the fine necklace and the flashing spear and courage blazing from his eyes—his son? The man could only stand and stare and stare, as if he could not believe his senses. And then a small yellow dog pulled himself over the gunwale of the canoe, fell at his master's feet. Uri. . . . Far

and gasping on the floor of the canoe. Uri, whimpering softly, crept to his master's side.

Off to the northeast a haze of light glowed up from the sea. Sometimes the lagoon of an atoll throws up just such a glow. It is the reflection of the lagoon upon the lower sky. Lifting his head, the boy watched it with dulled eyes, uncomprehending at first.

"*Te mori*," he whispered at last, his voice a thread of awe. "The lagoon-fire."

There came a whir and fury in the sky above, a beat of mighty wings: an albatross, edged with light, circled above the canoe. It swooped low, its gentle, questing eyes turned upon the boy and his dog. Then the bird lifted in its effortless flight, flew straight ahead and vanished into the lagoon-fire. And then Mafatu knew. Hikueru, his homeland, lay ahead. Kivi. . . .

A strangled cry broke from the boy. He shut his eyes tight and there was a taste of salt, wet upon his lips.

The crowd assembled upon the beach watched the small canoe slip through the reef-passage. It was a fine canoe, artfully built. The people thought at first that it was

those cool depths it seemed to him that he could see faces . . . his mother's, perhaps. . . . He dashed his hand across his eyes. Had the sun stricken him daft? Had he been touched by moon-madness? Then a wave of overpowering anger brought him to his knees: anger at this dark element, this sea, which would destroy him if it could. His voice was thick and hoarse, his throat bursting with rage.

"Moana, you Sea God!" he shouted violently. "*You!* You destroyed my mother. Always you have tried to destroy me. Fear of you has haunted my sleep. Fear of you turned my people against me. But now"—he choked; his hands gripped his throat to stop its hot burning—"now I no longer fear you, Sea!" His voice rose to a wild note. He sprang to his feet, flung back his head, spread wide his arms in defiance. "Do you hear me, Moana? I am not afraid of you! Destroy me—but I laugh at you. Do you hear? *I laugh!*"

His voice, cracked but triumphant, shattered the dead air. He sank back on his haunches, shaking with spasms of ragged laughter. It racked his body, left him spent

body that supported it. But he knew from the size of the dorsal that it must be a tiger-shark. . . . It began at length to play upon his nerves and set him jumpy. He scarcely dared now, at night, to lash his steering paddle and sleep.

The sail slatted and banged. The boy paddled through the long hours, paddled until the muscles of his arms and shoulders ached in agony and every sinew cried in protest. And at night, when darkness brought blessed release from the sun, there was always the Fishhook of Maui leading him on. But now when the boy looked up at the ancient constellation, doubt lay heavy on his heart. Hikueru—where was it? Where was it? The sea gave back no answer.

"Maui," the boy whispered, "have you deserted me? Have you looked into my heart and found me wanting?"

And suddenly, like the snapping of a string, he was overwhelmed with despair. Maui *had* deserted him. It was Moana, the Sea God's turn. The sea looked dark and cool and inviting. Little wavelets lapped and chuckled about the hull like beckoning hands. He looked overside. Deep down in

like friendly eyes, luring Mafatu on; and torch-fishes darted up from the lower depths while the black waters gleamed with strange lights. Then gradually as the canoe entered some other current of the sea, the wind slackened and diminished, this wind that had blown for him so long. Now the sail began to slat and bang in the dead air. The canoe drifted on the slow, majestic tide. The ceaseless heave and surge of the sea's dark breast lulled the boy to half-sleeping rest; the murmur of the waters playing about the prow sounded like his mother's reassuring voice.

Each sun as it rose seemed hotter than that of the preceding day. Now the face of the ocean was a disk of blazing copper. Masses of seaweed, heavy with the eggs of fishes, floated on the sluggish tide, seeming to clutch at the canoe, to hold it back from its destination. Hikueru, the Cloud of Islands—did they really exist? Were they not, like the chambered nautilus, only an iridescence dreamed by the sea? Sharks were beginning to appear—as they did always about a craft becalmed. One dorsal fin, larger than the others, followed the canoe in leisurely parallel; just far enough distant so that Mafatu could not see the

flagration like the death of a world. Night came and fled. Dawn rose in a burst of flame, and still Mafatu's canoe skimmed across the sea currents, now on this tack, now on that.

He was to learn in the hours to come that all days, all time, would be like that: hours of blasting heat, of shattering sunlight; nights of fitful respite and uneasy sleep. Only the sea and the sky, the sea and the sky. A bird now and then, a fish leaping from the sea, a boy in a frail canoe. That was all.

As one day dragged into another, Mafatu scanned the heavens for some hint of rain. Storm, anything would have been a welcome relief to this blasting monotony, to this limitless circle of sea. His store of *poi* vanished. The coconuts likewise. His water was being guarded, drop by drop. But there would come a moment when the last drop must be taken, and then. . . .

The season of storm was long past. The days, as they came, were cloudless and untroubled. Each day broke like a clap of thunder and night fell softly as a footfall. The sea was sparkling and benign. The sun's rays were unbroken in their violence. At night the Fishhook of Maui twinkled down

waters. Far off an albatross caught the light of gold on its wings. Was it Kivi? Mafatu could not tell. The wind held fresh and fair. The high island had vanished over the curve of the sea and the eaters-of-men had vanished with it. But now the great ocean current that had carried Mafatu so willingly away from Hikueru was set dead against him.

He put his little craft first on one tack, then on another. As the long hours passed, it seemed as if he were making no headway at all, even though the canoe still cut smartly through the water. There was a drift and pull that appeared to make a forward gain impossible. Was it Moana, the implacable one, trying to prevent Mafatu from returning to his homeland?

"Perhaps," the boy thought wearily, "Maui is not yet ready for me to return. Is there still a shadow of fear in my heart? Is that it?" He was tired now in every nerve and sinew, tired in the marrow of his bones, tired of struggle.

The long hours passed slowly while the sun climbed. Mafatu lashed the steering paddle and slept fitfully. Uri lay in the shadow of the sail. The sun sank in a con-

"Not yet, Moana," he muttered fiercely. "You haven't won. Not yet."

But with falling night the wind still held. Darkness rose up from the sea, enveloping the world. The stars came out clear and bright. The boy searched among them for some familiar constellation to steer by. Should it be *Mata Iki*—Little Eyes? Would *Mata Iki* lead him back safe to Hikueru? And then he saw, and knew: there, blazing bravely, were the three stars of the Fishhook of Maui. Maui—his sign. Those were his stars to steer by. They would lead him home. In that moment he was aware that the chanting of his pursuers had become fainter, steadily diminishing. At first he could not believe it. He listened intently. Yes—there was no doubt about it: as the breeze freshened, the sound grew fainter each passing moment.

The boy quenched his thirst, ate a scrap of *poi*, fought against sleep as the night waxed and waned.

By daybreak the chanting had ceased altogether. There was no sign of the canoes upon the broad expanse of the sea. The sunburst marched across the swinging

was employing every art and wile of sailing that he knew. As long as the wind held he was safe. He managed his little craft to perfection, drawing from it every grace of speed in flight.

He knew that with coming night the wind might drop, and then— He forced the thought from his mind. If the wind deserted him it would mean that Maui had deserted him, too. But the savages would never get him! It would be Moana, the Sea God's turn. The boy looked down into the blue depths overside and his smile was grim:

Vengeance powered their muscles. They were tireless. On they came.

The wind dropped. Imperceptibly at first. Sensing it, Mafatu whistled desperately for Maui. "Maui ē! Do not desert me," he prayed. "This last time—lend me your help."

Soon the black canoes were so close that the boy could see the shine of dark bodies, the glint of teeth, and flash of ornament. If the wind died, he was lost. . . . Closer, closer the canoes advanced. There were six of them, filled each with ten warriors. Some of them leaped to their feet, brandished their clubs, shouted at the boy across the water. They were a sight to quake the stoutest heart. With every second they were cutting down the distance which separated their canoes from Mafatu's.

Then the wind freshened. Just a puff, but enough. Under its impetus the little canoe skimmed ahead while the boy's heart gave an upward surge of thanks. Maui had heard his prayer and answered.

Day broke over the wide Pacific.

There were the six black canoes, paddles flashing, now gaining, now losing. The boy

prayer. Then he was caught in the rip-tide. The outrigger dashed through the passage, a chip on a torrent. The wind of the open sea rushed to greet it. The sail filled; the outrigger heeled over. Mafatu scrambled to windward to lend his weight for ballast. He was off! Homeward, homeward. . . .

Soon, rising above the reef thunder, the boy could hear a measured sound of savage chanting. They were after him!

Looking back over his shoulder, he could see the dark shapes of the canoes rounding the southern headland. Moonlight shone on half a hundred wet paddles as they dipped and rose to the rhythm of the chant. It was too dark to see the eaters-of-men themselves, but their wild song grew ever more savage as they advanced.

The breeze was almost dead aft. The crab-claw sail swelled smooth and taut, rigid as a block of silver against the sky. The little canoe, so artfully built, ran with swell and favoring wind, laying knots behind her. She was as fleet and gracile as the following gulls. But the eaters-of-men were strong paddlers with death in their hearts. Their *motu tabu* had been profaned by a stranger.

reached to lay hold of the outrigger. His black hand clutched the *purau* pole. The canoe slacked. Mafatu could see the gleam of bared teeth. The boy lifted the paddle and cracked it down. . . . With a groan the man dropped back into the water. The canoe, freed, skimmed out toward the barrier-reef.

The savages stopped, turned back toward shore. Then they were running back to the trail that led across the island, shouting to their fellows as they ran. Mafatu knew that it was only a question of minutes until the whole pack would be aroused and in pursuit. But he had the advantage of a head start and a light craft, while their canoes would have to beat around the southern point of the island before they could catch up with him. If only the breeze held. . . . Mafatu remembered then that the canoes he had seen drawn up on the beach had not been sailing canoes. There were strong arms to propel those black canoes, to overtake him if they could.

Mafatu's canoe, so slim and light, sped like a zephyr across the lagoon. The tide was on the ebb, churning in its race through the passage into the outer ocean. The boy gripped the steering paddle and offered up a

His canoe. If only he could reach it, shove it into the water before the savages overtook him. Then he would be safe. . . .

He knew this trail as he knew the back of his hand. That knowledge gave him an advantage. He could hear his pursuers, slipping, stumbling through the brush, shouting threats in a language strange to his ears. But there was no mistaking the meaning of their words.

On the boy dashed, fleet as an animal. Thorns and vines clutched at him. Once he tripped and sprawled headlong. But he was up and away in an instant. Through the trees he caught a glimpse of white beach and his heart surged. Then he was speeding across the sand, Uri at his heels.

The canoe was at the lagoon's edge. The boy fell upon the thwart, shoved the craft into the water. The logs under the stem rolled easily. In that second, the black men, yelling wildly, broke from the jungle and dashed across the beach. Mafatu was not a minute too soon. He leaped aboard and ran up the sail. The savages rushed after him into the shallows. A gust of wind filled the sail. It drew smartly. Now the men were swimming. One of them, in the lead,

conch shells, an eerie wailing, like the voices of souls lost in interstellar space.

Mafatu saw that the savages were armed with ironwood war clubs—clubs studded with sharks' teeth or barbed with the sting-ray's spike. Zigzags of paint streaked their bodies. And towering above all, the great stone idol looked down with sightless eyes, just as it had looked for untold centuries.

Mafatu, lying there on the ledge of basalt, watched the strange scene, powerless to move, and he felt Doom itself breathing chill upon his neck. He drew back from the edge of the cliff. He must flee! In that very instant he heard a crashing in the under-growth, not twenty yards away. A guttural shout ripped the darkness. The boy flung a desperate glance over his shoulder. Four black figures were tearing toward him through the jungle; he could see them now.

He turned and ran blindly down the trail whence he had come. Slipping, sliding, stumbling, his breath all but choking in his throat. He felt like a man drowning in ice-cold water. He felt as he had sometimes felt in dreams, fleeing on legs that were weighted. Only one thought gave him cour-age as he ran: his canoe, ready and waiting.

from hollowed logs a rhythm that was a summation of life, a testament of death. Uri crept close to his master's side, the hair ridging on his neck, his growl drowned in the thunder.

Now the boy could see into the cleared circle of the Sacred Place. Leaping fires lighted a scene that burned itself forever into his memory. Fires blazed against the basalt cliffs, spurts of flame that leaped and danced, showers of sparks borne off on the back of the night wind. Now a deep wild chanting rose above the booming of the drums. From his vantage point Mafatu saw six war canoes drawn up on the beach. Mighty canoes they were, with high-curving stems and decorations of white shell that caught the firelight in savage patterns. But what held the boy's eyes in awful trance were the figures, springing and leaping about the flames: figures as black as night's own face, darting, shifting, bounding toward the sky. The eaters-of-men. . . . Firelight glistened on their oiled bodies, on flashing spears and bristling decorations. High above the drums' tattoo rose the mournful note of the

was softly brilliant in the light of the waning moon. Any figure moving along the sand he could have seen instantly. He stopped, every nerve strung like a wire: the beach was deserted, the jungle silent and black with mystery. The boy rose to his feet. Swift as a shadow he turned into the undergrowth where the trail led to the high plateau. The jungle had never seemed so dark, so ominous with peril. The tormented roots of the *mapé* trees clutched at him. Lianas tripped him. Tree-ferns, ghostly in the half-light, rustled about him as he passed, their muted hush seeming to say: "Not yet, Mafatu, not yet." . . .

By the time he reached the plateau he was breathless and panting. He dropped to his knees and crawled forward, an inch or two at a time. One false move might spell destruction. But he had to know, he *had* to know. . . .

Thump-thump THUMP!

The measured booming grew louder with every inch that he advanced. Now it pounded in his ears, reverberated through his body, thrummed his nerves. Somewhere below in darkness, black hands drew forth

5

Homeward

A CHILL SWEAT broke out over Mafatu's body. He crouched there listening, unable for the moment to make a single movement. The rhythmic message that boomed across the mountain brought him a message of doom. Thump-thump THUMP! Thump-thump THUMP! It shivered along his nerves, setting his hair on edge.

Warily, moving with utmost caution, Mafatu crept out of the house. The beach

and faced it down. Surely that could be called courage.

When he lay down to sleep that night there was a profound thankfulness in his heart. "Tavana Nui," he whispered, "my father—I want you to be proud of me." . . .

He fell into a heavy, dreamless sleep.

Before dawn he was awakened by a sound of measured booming, like the beating of a supernatural drum. Thump-thump THUMP! Thump-thump THUMP! It rose above the thunder of the reef, solemn and majestic, filling the night with thunder.

Instantly awake, listening with every sense, Mafatu sat upright on the mats. Far out on the reef the seas burst and shot upward like sheeted ghosts in the moonlight. There it came again: Thump-thump THUMP! Thump-thump THUMP! Steady as a pulse, beating in the heart of darkness. . . .

And then Mafatu knew. The eaters-of-men had come.

As Mafatu waited for his supper to cook, he set about preparing for his homeward journey; he would start at daybreak with the ebbing tide. He would be ready. He filled bamboo containers with fresh water, sealed them with leaves that were gummed into place, water-tight and secure. Then he stored them carefully in the canoe. He prepared a *poi* of bananas and sealed it, likewise, into containers; there it would ferment and sour and become delicious to the taste. Then he picked a score or more of green drinking nuts and flung them into the canoe. And as he trotted back and forth across the beach and his supper steamed on the fire, one thought alone, like an insistent drum beat, echoed in the boy's heart: "Tomorrow I shall start home! Tomorrow, tomorrow!"

Never again need he hang his head before his people. He had fought the sea for life and won. He had sustained himself by his own wits and skill. He had faced loneliness and danger and death, if not without flinching, at least with courage. He had been, sometimes, deeply afraid, but he had faced fear

illuminated fishes moved and had their being: meteors, galaxies, constellations under the sea. The boy saw a line of light, narrow as a blade, as the rare *pala* flashed away in its everlasting quest. A sand shark, phosphorescent ghost, darted after the *pala*—seized it in a swirl of luminous mist. The mist faded slowly. It was blood. Mysterious life forces were completing their cycle in those dark depths, even as on the earth and in the air above. This sea—no more to be feared than earth or air: only another element for man to conquer. And he, Mafatu, had killed the *feke*. *Auê te auê!*

As he dipped his paddle with a swinging rhythm, the rhythm of his thoughts swung in unison: "Tomorrow I shall start home! Tomorrow, tomorrow! *Aiâ!*"

The very thought of it set him aquiver. "Tomorrow, tomorrow!" he had been here so long. . . .

He dragged the canoe up on the beach, placed the logs under the curving stem so that he might launch it easily on the morrow. He would never need to climb the high plateau to the lookout again. Let the eaters-of-men come!

With sharkline and hook the boy fished up the *fekē's* body. As he dragged it into the canoe one of the tentacles brushed his ankle. Its touch was clammy and of a deathly chill. Mafatu shuddered and shrank away. He had eaten squid and small octopi ever since he was born, but he knew that he could not have touched a mouthful of this monster. He raised his spear and plunged it again and again into the body of his foe, shouting aloud a savage paean of triumph. A thousand years of warrior-heritage sounded in his cry.

Once more Maui had protected him! What to do with the *fekē*? The boy decided that he would cut off the tentacles; they would dry and shrink, but they would be still of prodigious size, and the people of Hikueru would say: "See, Mafatu killed the *fekē* single-handed. *Auē te auē!*"

Dusk, almost in an instant, deepened into night. As Mafatu turned the nose of his canoe toward shore, the first stars were appearing, bright and close and friendly. There was the Southern Cross, pointing toward the end of the world. . . . The lagoon was a black mirror dusted with starshine. Far below in the dark waters,

out his venom. There in submarine gloom a boy fought for his life with the most dreaded monster of the deep. He could feel the sucking pressure of those terrible tentacles. . . . His wind was almost gone.

Blindly Mafatu stabbed again, this time for the other eye. The blow, so wildly driven, went true. The terrible grip relaxed, slacked. The tentacles grew limp. Then Mafatu was springing upward, upward, drawn up toward light and air and life.

When he reached the canoe he had hardly enough strength to cling to the gunwale. But cling he did, his breath coming in tearing gasps. Uri, beside himself, dashed from one end of the canoe to the other, crying piteously. Slowly strength returned to the boy's limbs, warmth to his chilled soul. He dragged himself into the canoe and collapsed on the floor. He lay there, as in a trance, for what seemed an eternity.

The sun had set. Dusk was rising from the surface of the sea. Mafatu struggled upright and peered cautiously over the side of the canoe. The inky water had cleared. Down there, forty feet below, the octopus lay like a broken shadow. The white cups of its tentacles gleamed dully in the watery gloom.

across and taller than he: its open lips waiting to snap shut upon fish or man. Green fronds waved gently as if in some submarine wind. A shadow moved above the boy's head and he glanced upward in alarm: only a sand shark cruising harmlessly. . . . An eel, like a cold waving ribbon, touched his leg and was gone.

The knife—there it lay. How sharp and bright it looked. Now the boy's hands were upon it. He seized it and sprang upward toward the light.

In that second a whiplash shot out from a cavern at his back: a lash like a length of rubber hose. The boy caught the flash of vacuum cups that lined its under surface. Panic stabbed him. The *feke*—the octopus! Another lash whipped forth and encircled his waist. It drew taut. Then the octopus came forth from its den to face and kill its prey.

Mafatu saw a purplish globe of body, eyes baleful and fixed as fate; a parrot-mouth, cruel and beaked, that worked and wabbled. . . . Another whiplash encircled the boy's leg. The knife— Desperately Mafatu stabbed for one of the eyes. Then darkness clouded the water as the octopus siphoned

use! The long hours that had gone into the making of the knife. . . . It had saved Uri's life, too. And now Uri, in the bow of the canoe, was looking at his master with eyes so puzzled and true.

Mafatu drew a deep breath. How could he abandon his knife? Would Maui (the thought chilled him) think him a coward? Was he still Mafatu, the Boy Who Was Afraid?

He leaped to his feet, gave a brave hitch to his *pareu*. Then he was overside in the water. He clung for a moment to the gunwale, breathing deeply. Inhaling, then releasing the air in a long-drawn whistle, he prepared his lungs for the pressure of the depths. Many times he had seen the pearl divers do it. In the canoe lay a coral weight fastened to a length of sennit. Mafatu took this weight and held the cord in his toes. With a final deep breath he descended feet-first, allowing the weight to pull him downward. At about twenty feet he released the weight, turned over, and swam for the bottom.

Here the water was cool and green. The sunlight filtered from above in long, oblique bands. Painted fishes fled before him. He saw a giant *pahua*, a clam shell, five feet

that fastened his knife about his neck caught on an end of bamboo. The trap slipped. The cord snapped. The knife fell into the water.

With dismay the boy watched it descend. It spiraled rapidly, catching the sunlight as it dropped down, down to the sandy bottom. And there it lay, just under the edge of a branching staghorn. Mafatu eyed it uncertainly. His knife—the knife he had labored so hard to shape. . . . He knew what he ought to do: he should dive and retrieve it. To make another knife so fine would take days. Without it he was seriously handicapped. He *must* get his knife! But . . .

The reef-wall looked dark and forbidding in the fading light. Its black holes were the home of the giant *feke*—the octopus. . . . The boy drew back in sudden panic. He had never dived as deep as this. It might be even deeper than he thought, for the clarity of the water confused all scale of distance. The knife looked so very near, and yet . . . There it lay, gleaming palely.

The boy gazed down at it with longing. He remembered the morning he had found the whale's skeleton; the first one be had ever seen. Surely Maui, God of the Fishermen, had sent the whale there to die for Mafatu's

tiny mullet, miniature arrowheads—the whole school scarcely larger than a child's hand. A conger eel drew its ugly head back within a shadowy cavern.

Here beside the wall of reef Mafatu's bamboo fish trap hung suspended; before he returned to shore he would empty the trap. It had been undisturbed since the hammerhead was killed, and each day had yielded up a good supply of mullet or crayfish or lobsters. Here the wall of living coral descended to the lagoon floor. Its sides were pierced with caves of darkness whose mystery the boy felt no desire to explore. Far below, perhaps forty feet, the sandy floor of the lagoon was clear and green in the dappled light. A parrot fish emerged from the gloom, nibbled at Mafatu's bait, then vanished.

"*Auê!* These fish must be well fed. My piece of crab meat does not tempt them."

The boy decided to give it up and content himself with the fish in the bamboo trap. He leaned over the gunwale and pulled it up out of water. Through the openings in the cage he could see three lobsters, blue-green and fat. What luck! But as he dragged the heavy, wet trap over the gunwale, the fiber cord

The reef—this was a part of his heritage. The sea, at last, was as much his element as the land.

The realization flooded through him in a warm tide of content. He lowered his fishline, fastened it to the mid-thwart, and looked deep down into the clear water. He saw a scarlet rock-cod materialize, hang in the shadow of the canoe, motionless save for the slight movement of its gills. With a sudden flip of the tail it vanished.

How fantastic was that undersea world! The boy saw branching staghorn corals, as large as trees, through which jellyfishes floated like a film of fog. He saw shoals of

The boy skirted along the edge of the reef, lowered his sail, and dropped overboard the lump of coral which served as anchor. Then he took out his fishline and baited the hook with a piece of crab meat. He wanted to enjoy to the full this new sensation of confidence in himself, this freedom from the sea's threat. He looked back at the land fondly, but without longing. The high peak, purple in the waning light, stood somber against the sky. The valleys were shadowed with mystery. All these weeks he had lived close to this island and been grateful for its bounty. But he had been born on an atoll—a low island—and all his life had been spent in the spaciousness of open sea and wind-swept palms. There was something gloomy and oppressive in this high island.

speed. It was late afternoon and the sun was setting in a blaze of glory, but the boy was reluctant to turn back. He knew that he should have climbed to the lookout that morning. This was the first day he had neglected that duty. But the temptation to complete his canoe had been too great. Tomorrow at daybreak he would climb the plateau for the last time. And then—and then Hikueru!

As the little craft skimmed out toward the barrier-reef, the thunder of the surf increased in volume to an overwhelming sound. Waves, born far south in the Antarctic ice fields—the home of all waves—broke their backs upon this coral rampart. Gathering far out, they charged the reef: sea horses with flinging manes of foam. The surf shot skyward and above its mist sea gulls swooped and darted. The reef thunder no longer filled Mafatu with unease. He had lived too close to it these past weeks. Out here, half a mile from shore, detached from all security of the land, he had come to believe that at last he had established a truce with Moana, the Sea God. His skill against the ocean's might.

it with shining eyes. He could hardly believe that all these weeks of labor were at an end. Suddenly he was quiet. With lifted head he offered up the prayer with which all ships were launched at Hikueru:

"Taaroa, Mighty One!
My thanks to you
In this task completed.
Guide it on your back
To safe harbor.
Taaroa, *é!*"

The boy leaped into the stern, picked up the paddle, and ran up the sail. Uri sprang into the bow, yelping for very joy. Kivi sailed far overhead on widespread wings. The breeze caught the sail, swelled it to a golden curve. The outrigger leaned at a sharp angle and sped ahead toward the distant reef. Spray flew back from the prow and Mafatu's heart beat high. He let out the sheet, wrapped the sennit rope around his foot, and gripped the steering paddle with both hands. He was filled with pride in his canoe. Never had he been as happy as in this moment.

Out toward the black reef, closer and closer the canoe skimmed on a wide arc of

Mafatu slipped the necklace about his throat, and he could fairly feel its magic charging him with strength! He pulled the oven stones away from the *umu*, and there lay the pig, golden, glowing, done to a turn. Rich juices ran in little rivulets down its sides. And as Mafatu ate, one thought alone filled his mind, overshadowing even his enjoyment of this rare feast: soon, soon now he would be ready. He had killed the *ma'o*. The *puaa*, too. His canoe would soon be completed. And then—then he would return to Hikueru!

The canoe was finished.

Mafatu lashed the tough *purau* outrigger into place with hands that trembled. The woven sail was complete and ready; the rigging strong as wire. There—it was all over! The boy could hardly wait to get his craft into water.

Placing logs under the curving stem, he gave a shove and a push. The canoe stirred, moved forward, quick with life. Another shove and the craft slid into the lagoon. And there it floated lightly, easily as a gull poised for flight. Mafatu stood back and surveyed

dozens of them—to hold in the steam and to allow the pork to cook through slowly. Uri leaped about, sniffing at the delicious odors, barking his delight. *Pork!* After weeks of fish and fish and fish, how good it would taste! Besides, fish was not good for dogs. Too many bones. Kivi, no meat eater, looked on calmly, wondering what all this disturbance was about; the bird was content with a coconut that Mafatu split open for him, that he might join in the feast.

Mafatu's mouth fairly watered in anticipation. But even as he settled back to await the feast, his hands were busy: the sun gleamed brightly on the curving tusks that already he was making into a necklace. They formed almost a complete circle and were as white as bleached coral. *Aué!* How men's tongues would chatter when they saw this fine necklace! Even Grandfather's had been no finer.

A strange picture on that lonely beach under the palms: a pig roasting on a fire; a boy lean and brown and whip-strong, making a boar's-tooth necklace; a prancing yellow dog; a calm, wide-winged albatross pecking at a coconut.

for this I saved you from the teeth of the *ma'o?* I've a mind not to give you one mouthful of *puaa.*"

Uri hung his head and dropped his tail, then brightened as Mafatu laughed joyously. "Foolish one! Here—drag your share."

The boy made a rude sled of bamboo and loaded the heavy animal onto it. Then he hitched a stout liana about Uri's neck and the dog threw his weight into the task. The two started home in triumph with their burden, Mafatu singing at the top of his lungs a lusty song of blood and battle. He was all Polynesian now, charged with the ancient fierceness of his race. Victory coursed like fire through his veins. There was nothing he would not have dared! Nothing he feared! *Aià*, but life was good!

When they reached the camp site, Mafatu built up a roaring fire and set a pile of stones to heat. While the stones were heating, the boy cleaned the pig at the water's edge, then stuffed its empty belly with succulent *ti* leaves and red bananas. When at last the oven stones were white and smoking, Mafatu dragged the pig back to the fire and rolled it upon the hot *umu*. Then he covered it with layer upon layer of plantain leaves—

run. Then he drew a deep breath and shouted out a challenge:

"*Puaa viri!* Wild pig! I, Mafatu, have come to kill you!"

The boar charged. Over the ground it tore. Foam flew back from its tusks. The boy braced himself. He met the charge with a perfectly timed thrust of the spear. The boar impaled itself, shoulder-deep, upon the spearhead.

Mafatu was thrown off balance, sent spinning headlong. Over and over he rolled. He leaped to his feet in a panic, defenseless. But the boar toppled, gave a convulsive shudder, lay still.

Mafatu was struck dumb. He had killed a wild pig! For a second he could not grasp the wonderful truth of it. Then he leaped wildly into the air, shouting: "*Auê te auê!* I have killed the *puaa!* Do you hear me, Tavana Nui? I, your son, have killed a boar! Ho! Ha!"

Uri came leaping out of the jungle and barked himself hoarse at sight of the pig.

"A fine one you are!" Mafatu joked at his dog. "Where were you when I needed you? Off chasing butterflies, that's you! Was it

"Today I will not go," the boy thought, as his adze whirred and chipped. "It takes too long."

Then he sighed and laid down the adze. Caution was the better part of wisdom. He picked up his shining spear with its new shaft and turned toward the trail that led to the high plateau. Uri leaped ahead, his nose keen-pointed to the ground.

This day as Mafatu climbed the rough trail through the jungle, he was preoccupied, lost in his thoughts. His mind was not in this business at all: he was thinking about the rigging of his canoe, planning how he could strengthen it here, tighten it there. Some instinct of danger must have made him pause, warning him to beware. There it was—just a rustle in the undergrowth. Scarcely louder than an insect's hum. The boy drew up tense and listening. Uri had dashed off on some wild-goose chase of his own. The boy stood rooted, alert. He saw it then: a wild boar with lowered head. Eyes red with hate. The flash of its wicked tusks.

The animal pawed the ground suddenly. Its grunting snort broke the stillness. Mafatu knew a blind impulse to turn and

made of pukatea, straight and true; a sail
woven from pandanus. And there was rig-
ging to be made of sennit, tough and strong
as wire. The craft would have been finished
sooner if only there were not so many things
to interfere. Every day, for example, Mafatu
climbed the plateau to his lookout. He had
not missed one day since he arrived at the
island. He knew that when the eaters-of-
men came they would sail by day; they
would have to beat against the prevailing
wind. It would take them many hours to
come from Smoking Island. Mafatu would
be able to see them from his lookout long
before they arrived. If only he could be
ready before they came! He must, he must!
But that trip up the mountain every day took
so much precious time. His canoe would
have been finished days ago, but for that.

in—other sharks, tearing the wounded hammerhead to pieces.

Uri—where was he? Mafatu saw his dog then. Uri was trying to pull himself up on the raft. Mafatu seized him by the scruff and dragged him up to safety. Then he caught his dog to him and hugged him close, talking to him foolishly. Uri yelped for joy and licked his master's cheek.

It wasn't until Mafatu reached shore that he realized what he had done. He had killed the *ma'o* with his own hand, with naught but a bone knife. He could never have done it for himself. Fear would have robbed his arm of all strength. He had done it for Uri, his dog. And he felt suddenly humble, with gratitude.

Now the adze was completed. Thus the canoe, too, was beginning to take finished shape. It was fifteen feet long, three feet deep, and scarcely a foot wide. Often as he worked, the boy would pause to stand off and admire his craft. It was a beautiful canoe! How proud his father would be. . . . Alas that it was such slow work.

When the hull had been hollowed out, it must be smoothed off with the adze and caulked with *artu* gum. Then a mast must be

74

large wave sheeted across the reef. At that second the dog's shift in weight tipped the raft at a perilous angle. With a helpless yelp, Uri slid off into the water. Mafatu sprang to catch him but he was too late.

Instantly the hammerhead whipped about. The wave slewed the raft away. Uri, swimming frantically, tried to regain it. There was desperation in the brown eyes— the puzzled eyes so faithful and true. Mafatu strained forward. His dog. His companion. . . . The hammerhead was moving in slowly. A mighty rage stormed through the boy. He gripped his knife. Then he was over the side in a clean-curving dive.

Mafatu came up under his enemy. The shark spun about. Its rough hide scraped the flesh from the boy's shoulder. In that instant Mafatu stabbed. Deep, deep into the white belly. There was a terrific impact. Water lashed to foam. Stunned, gasping, the boy fought for life and air.

It seemed that he would never reach the surface. *Auê*, his lungs would burst! . . . At last his head broke water. Putting his face to the surface, he saw the great shark turn over, fathoms deep. Blood flowed from the wound in its belly. Instantly gray shapes rushed

was like a triangle of black basalt, making a little furrow in the water as it passed.

"*Aiá*, Ma'o!" the boy shouted roughly, trying to bolster up his courage. "I have my knife today, see! Coward who robs traps—catch your own fish!"

The hammerhead approached the raft in leisurely fashion; it rolled over slightly, and its gaping jaws seemed to curve in a yawning grin. Uri ran to the edge of the raft, barking furiously; the hair on the dog's neck stood up in a bristling ridge. The shark, unconcerned, moved away. Then with a whip of its powerful tail it rushed at the bamboo fish trap and seized it in its jaws. Mafatu was struck dumb. The hammerhead shook the trap as a terrier might shake a rat. The boy watched, fascinated, unable to make a move. He saw the muscles work in the fish's neck as the great tail thrashed the water to fury. The trap splintered into bits, while the fish within escaped only to vanish into the shark's mouth. Mafatu was filled with impotent rage. The hours he had spent making that trap— But all he could do was shout threats at his enemy.

Uri was running from one side of the raft to the other, furious with excitement. A

"Wait, you!" the boy threatened darkly, shaking his fist at the *ma'o*. "Wait until I have my knife! You will not be so brave then, Ma'o. You will run away when you see it flash."

But the morning that the knife was finished, Mafatu did not feel so brave as he would have liked. He hoped he would never see the hammerhead again. Paddling out to the distant reef, he glanced down from time to time at the long-bladed knife where it hung about his neck by a cord of sennit. It wasn't, after all, such a formidable weapon. It was only a knife made by a boy from a whale's rib.

Uri sat on the edge of the raft, sniffing at the wind. Mafatu always took his dog along, for Uri howled unmercifully if he were left behind. And Mafatu had come to rely upon the companionship of the little yellow dog. The boy talked with the animal as if he were another person, consulting with him, arguing, playing when there was time for play. They were very close, these two.

This morning as they approached the spot where the fish trap was anchored, Mafatu saw the polished dorsal of the hated hammerhead circling slowly in the water. It

Fishing with a line took too long when you were working against time. Mafatu could not afford to have his trap robbed. Twice it had been broken into, the stout bamboos crushed and the contents eaten. It was the work either of a shark or of an octopus. That was certain. No other fish was strong enough to snap the tough bamboo.

Mafatu's mouth was set in a grim line as he worked away on his knife. That old hammerhead—undoubtedly *he* was the thief! Mafatu had come to recognize him; for every day when the boy went out with his trap, that shark, larger than all the others, was circling around, wary and watchful. The other sharks seemed to treat the hammerhead with deference.

Hunger alone drove Mafatu out to the reef to set his trap. He knew that if he was to maintain strength to accomplish all that lay ahead he must have fish to add to his diet of fruit. But often as he set his trap far out by the barrier-reef, the hammerhead would approach, roll over slightly in passing, and the cold gleam of its eye filled Mafatu with dread and anger.

shouldered himself. The other Uri dragged behind him. And thus they returned to the camp site, weary, but filled with elation. Even the dog seemed to have some understanding of what this discovery meant; or if not, he was at least infected with his master's high spirits. He leaped about like a sportive puppy, yapping until he was hoarse.

Now began the long process of grinding the knife and the ax. Hour after long hour, squatting before a slab of basalt, Mafatu worked and worked, until his hands were raw and blistered and the sweat ran down into his eyes. The knife emerged first, since that was the most imperative. Its blade was ten inches long, its handle a knob of joint. It was sharp enough to cut the fronds of coconut trees, to slice off the end of a green nut. *Ai*, but it was a splendid knife! All Mafatu's skill went into it. It would be a fine weapon as well, the boy thought grimly, as he ground it down to a sharp point. Some sea robber had been breaking into his bamboo trap and he was going to find out who the culprit was! Probably that old hammerhead shark who was always cruising around. . . . Just as if he owned the lagoon!

breezes of the lagoon. It was a trim little house and he was proud of it. A roll of woven mats lay on the floor; there was a shelf in the wall with three bowls cut from coconut shells; bone fishhooks dangled from a peg; there was a coil of tough sennit, many feet long; an extra *pareu* of tapa waterproofed with gum of the *artu* tree, for wet weather. All day long the wind played through the openings in the bamboo walls and at night lizards scurried through the thatch with soft rustlings.

One morning, wandering far down the beach, Mafatu came upon a sheltered cove. His heart gave a leap of joy; for there, white-gleaming in the sun, was all that remained of the skeleton of a whale. It might not have meant very much to you or to me; but to Mafatu it meant knives and fishhooks galore, splintered bone for darts and spears, a shoulder blade for an ax. It was a veritable treasure trove. The boy leaped up and down in his excitement. "Uri!" he shouted. "We're rich! Come—help me drag these bones home!"

His hands seemed all thumbs in his eagerness; he tied as many bones as he could manage into two bundles. One bundle he

nesting along the shore to be chased and put to rout; and wild goats and pigs in the mountains to make life exciting enough for any dog.

Mafatu had discovered a mulberry tree. He stripped off the bark and removed the inner white lining. Then he wet the fiber and laid it upon a flat stone and set about beating it with a stick of wood. The fiber spread and grew thinner under the persistent beating. The boy added another strip, wet it, and beat it into the first one; then another and another. Soon he had a yard of "cloth" to serve as a *pareu*. It was soft and white, and now at last he was clothed.

"Before I go home I will make a dye of *ava* and paint a fine design on my *pareu*," the boy promised himself. "I must not go back ill-clothed and empty-handed. Men must know that I have conquered the sea, and made the land serve me as well."

The days passed in a multitude of tasks that kept Mafatu busy from dawn till dark. His lean-to grew into a three-sided house with bamboo walls and a thatch of palm leaves. The fourth wall was open to the

necklace was almost as important as a canoe! For by that token men would know his strength and courage. When the day came that he should leave this high island, he would sail to the north and east. Somewhere in that quarter lay the Cloud of Islands; the great Tuamotu Archipelago which extends across a thousand miles of ocean and ten degrees of latitude. Within those reef-spiked channels floated Hikueru, his homeland. . . . There was no doubt in his mind that he would find it; for Maui, who had led him safe to this shore, would someday guide him home again. But first, Mafatu knew, he must prove himself worthy. Men should never again call him Mafatu, the Boy Who Was Afraid. And Tavana Nui should say with pride: "Here is my son, come home from the sea."

Kivi, the albatross, came and went on his mysterious errands, emerging out of blue space, vanishing into it again. At sundown, regularly, the white bird came wheeling and circling, to alight clumsily on the beach almost at Mafatu's side, while Uri pranced about and greeted his friend after his own fashion. As for Uri, he was having the time of his life; for there were countless sea birds

the fire ate deeper into the trunk, the tree began to lean under the boy's weight. With a snap and a crash it fell across the sand. As it fell, Mafatu leaped free of the branches, as nimbly as a cat.

"That's enough for today, Uri," he decided. "Tomorrow we'll build our fires down the trunk and start burning it out. When the eaters-of-men come, we will be ready!"

In the meantime there were many other things to do: a fish trap of bamboo, a net of sennit, a fishhook, too, if only he could find some bone. And while the canoe was building, how could Mafatu get out to the distant reef to set his trap, unless first he made a raft of bamboo?

The boy decided that the raft was of first importance. He chose a score or more of fine bamboos as large around as his arm, felling them by fire; then he lashed them together with strips of *purau* bark, making a sturdy raft of two thicknesses. It would serve him well until his canoe should be finished.

As he worked, his mind returned again and again to the wild pig he was determined to kill. How could he go back to Hikueru without a boar's-tooth necklace? Why, that

hand. Woe unto the small boy who let the family fires go out!

While his breakfast roasted in the coals, the boy cleared the brush away from the base of the great tamanu. There was no wood better for canoe building than this. It was tough, durable, yet buoyant in the water. Mafatu could fell his tree by fire, and burn it out, too. Later he would grind an adze out of basalt for the finished work. The adze would take a long time, but he had made them often in Hikueru and he knew just how to go about it. The boy was beginning to realize that the hours he had spent fashioning utensils were to stand him now in good stead. Nets and knives and sharkline, implements and shell fishhooks—he knew how to make them all. How he had hated those tasks in Hikueru! He was quick and clever with his hands, and now he was grateful for the skill which was his.

The fire crackled and snapped about the base of the tamanu tree. When at length it had eaten well into the trunk, Mafatu climbed aloft and crept cautiously out upon a large branch that overhung the beach. Then taking firm hold of the branches above his head, he began to jump up and down. As

4

Drums

THE VERY next morning Mafatu set about building his canoe. He had banked his fire the night before in the natural shelter of a cave and he resolved never to let the sparks die out. For it was altogether too difficult to make fire with the firestick, and it required too much time. In Hikueru, for that reason, the fires were always kept burning, and it was the special charge of the younger members of a family to see that fuel was ever at

sound of his greedy munching. Never had food tasted so good!

He ate and ate, until his stomach could hold no more. Then a great weariness stole through his limbs. With a sigh of deep content he sank back upon the cool sand and closed his eyes. Uri snuggled close, warm and friendly against his side. Overhead, the plaited fronds of the palms formed a snug shelter. "Someday," the boy thought drowsily, "I will make my roof water-tight, but not yet." . . .

The tide was running out of the lagoon with the sunset in a whispering hum of sound, like the reassuring hush of a mother to her child. Mafatu lay there under his lean-to, relaxed in every nerve. He had fire, food, shelter. He had faced Moana, the Sea God. He had dared the sacred marae of the eaters-of-men to win his spear. There was a new-found confidence singing in his heart. He had found a new belief in himself.

At peace with himself and his world, he slept.

then he will say: '*Aiã!* Is it possible—this fine canoe?'"

The fire blazed brightly and the breadfruit gave forth a tempting odor. Jets of vapor rose from the steaming bananas. Uri snuggled down at his master's side and Mafatu hugged the dog close. "*Ai*, there's food for you, too, my brother," he cried. "And tomorrow we shall have fish. See my spear, how bravely it flashes! I will make a new shaft for it, and then I shall kill the *ma'o*, the shark! And the wild pig, too. *Ai*, Uri, you will see!"

The boy pulled the breadfruit from the fire and broke it in half. Its white, flaky meat was packed with sustenance; a little like bread in taste, a trifle like potato, better than either. The boy gave half of it to Uri. The dog circled around, sniffing longingly as he waited for his dinner to cool. The bananas popped from their skins while their juices bubbled in the heat. Deftly Mafatu slid them out of the fire, transferred them to a broad leaf that served as a plate. Then, unmindful of burned fingers, he fell upon the hot food. He was famished. It seemed that he had never been so hungry in his whole life, and for a time there was only the

down to a fine cutting edge. We'll soon have a good knife."

Uri thought so, too; that was evident from the motion of his tireless tail.

"Perhaps," Mafatu thought suddenly, "I might find a knife at the Sacred Place of the eaters-of-men." Then he shook his head. "No! I shall never return to that terrible marae *tabu!*"

But each day, he resolved, he would climb to the high plateau, his lookout to the southwest. Someday, someday the eaters-of-men would come back. Before long, perhaps. . . . Mafatu was certain of it, as he was certain that the sun shone and was warm. It was inevitable. And when the black men came, he would have to leave, or die.

The boy looked fondly at the tamanu tree that he had marked already for his canoe. "Tomorrow I shall start work upon it," he promised. "And *aiá*, what a canoe I shall build! Deep and strong, but light. An outrigger as swift and powerful as a shark's tail. And when my father sees it he will say: '*Aué*, my son, but they build fine canoes in that island you have come from.' And then I will say: 'But I built it myself, my father.' And

would not think of those things. He would not remember—if he could help it. Now he would have cooked food, warmth when he was wet through, and the first implement with which to set about building his canoe. Tomorrow he would fell a tree by fire and start to work upon his dugout. He would make fire only by day, and then not too large a blaze, for its glow might be visible on the distant island of the eaters-of-men. Thus he forced his thoughts into safer channels.

He threw breadfruit into the fire. There it would stay until it was charred black and thoroughly cooked. Next, he laid half a dozen *fei*—wild bananas—in the coals, covering them with wet leaves so that slowly they would steam through. While his dinner was cooking, he had yet another task to do. He set about it at once: he tore off half a dozen fronds of a low-growing coconut tree; then he began to plait them into flat, matlike screens. These, made into a lean-to, would be his shelter against wind and storm. Several layers of them would be water-tight.

"If only I had my knife, Uri," the boy groaned. "How simple it would be! I must search the lagoon for a *pahua* shell. It grinds

and forth upon the hard surface. A groove commenced to form. A little pile of wood dust gathered at one extremity of the groove. Now the boy's hands were moving faster and faster. Back and forth they flew, while perspiration stood out on his forehead and his breath came in gasps. At last he was rewarded. A wisp of smoke drifted upward from the wood dust. A glow. . . . The boy seized a few twigs and, cupping his hands with great care, blew upon the spark. A flame burst into life.

He had made fire.

Weary with the exertion, Mafatu settled back on his haunches and watched the leaping flame. *Fire!* His arms and back ached unmercifully; the wound in his leg throbbed. But the warm glow of flame comforted him. It made him think of home, of food and warmth and companionship; of faces around the evening circle; of the drone of old men's voices, telling their endless tales of daring. And he was swept by a sudden wave of loneliness, a longing for the sound of his father's deep voice. . . . He shut his lips tight and fought it back, then leaped to his feet and set about his small tasks with a great show of business. He

And as he set about searching for a firestick he sang a song at the top of his young lungs. It was a brave song of Taaroa, the hero-god who rose out of the sea to slay the enemies of his people:

"Taaroa puai e!
Hiti raa no te moana,
Horoa te puai ia'u,
Taaroa e!"

His voice rose clear and strong across the jungle stillness. Sea birds and parakeets hushed their cries to listen to the strange, intruding sound.

At last Mafatu found a firestick to his liking: a piece of hard wood, bone-dry, as large around as his forearm. Then he searched for a smaller piece of the same wood. Propping the larger piece against a rock, he squatted on his haunches before it and gripped the small sliver in both hands. He began to move it back and forth, back

56

At last he gained the plateau—drew up breathless and panting. The spear flashed in his hand. He looked down at it in wonder. Nothing had happened to him! He had touched the marae, dislodged a bone on the place *tabu*, yet still he lived. The sun shone. The sky arched blue. Nothing was changed. And this spear, *auê!* But it was a beautiful one! It was worth all the risk. Those eaters-of-men had wrought well. Now he, Mafatu, could kill the wild pig! Now he could defend himself against attack.

But, most important of all, he knew that he had won a great victory over himself. He had forced himself to do something that he dreaded, something that took every ounce of his will. The taste of victory salted his lips. He threw out his chest and cried: "It is you, Maui, who have helped me! My thanks, my thanks to you!" Uri leaped and pranced with excitement at his master's side.

Back once more at his own shelter, Mafatu kept looking and looking at the spear. Happiness flooded through him in warm tides. It was not so much the possession of a spear. No . . . it was the fact that he had touched the marae. That took courage, *ai*, courage!

shadows moving among the ferns and hear the phantom whisper of voices. But he edged forward, poised for instant flight. There—he was so close to the idol that he could have touched it. He reached out his hand. It took every ounce of will. The spearhead glistened brightly. . . . His fingers closed about it, tightened. The towering idol cast a shadow of darkness across the green earth. Quickly the boy drew the spearhead toward him. But in moving it he dislodged a bone. It fell at his feet. Its touch was deathly cold. Mafatu gasped. Then he whipped about and was running, running. But he still gripped the spearhead in his fist.

He scrambled back up the path whence he had come. His heart was hammering, his leg stiff and sore. The twisted roots of the *mapé* trees reached for him as he fled. Never had they seemed so much like grasping fingers. The giant tree-ferns turned the jungle to an eerie gloom. The boy was filled with unaccountable dread. His limbs seemed weighted. When at last he reached the base of the lava slide, he turned at right angles and fought his way upward, hauling, pulling himself by vines.

hear Moana saying: "Someday, someday, Mafatu, I will claim you." . . .

It was evident that the savages had been here recently, for the piles of ashes rested undisturbed by wind and storm. The cleared circle seemed to hold its breath, locked in a supernatural silence.

As the boy paused, irresolute, looking up at the towering marae, his eye was caught and held by a gleam of light and his heart gave a mighty leap. For he saw that a spearhead lay on the sacred platform. Finely ground, sharp-edged; a spear for food, a weapon against attack. Dare he take it? It might mean death. . . . His heart pounded. He moved one foot forward. His hands were damp and cold. The flashing spearhead winked back at him like an evil eye. The boy's limbs turned to water. For a second he was powerless to move. In that moment had a score of black men leaped forth from the jungle he could not have stirred or cried. He fought himself for control. Taking a deep breath, he whispered: "It is you, Maui, who have led me to this island. I know it. Do not forsake me now!"

Almost it seemed as if he could see dark

Scarcely daring to breathe, he advanced a step. Then he drew up short. Around the base of the idol he saw piles of bones, charred, but not old. The platform was strewn with them. Bones too large for dogs, too large for pigs. And then Mafatu understood. His heart congealed. This was a *motu tabu*, a Forbidden Island. Here the eaters-of-men made their terrible sacrifices to the *Varua Ino*.

Mafatu stood rooted, unable to advance, powerless to flee. Uri slunk at his side growling low, the hair on his neck lifting stiffly. Involuntarily the boy looked up; through an opening in the trees the cloud-cone of Smoking Island floated on a wine-colored sea. . . . Could that distant island be the home of the savages who used this *motu tabu* for their Place of Sacrifice? Was it here that they came in their black canoes to turn the night hideous with their drums and ceremonies and leaping fires? Mafatu shivered. He had believed that Maui, God of the Fishermen, had led him safe to this island. But perhaps after all it had been only a cruel prank of Moana, the Sea God, angry at having been cheated. The boy seemed to

plateau, for at that moment his eye fell upon a broad, well-defined path leading down through the jungle.

"*Aiâ!*" he exclaimed. "This fine trail was never made by the feet of wild pigs!"

The boy stood there irresolute and uncertain. Some premonition of danger kept him poised, alert, and wary. He was half tempted to retrace his way and explore no further. Then an overpowering desire to know who had made that path urged him onward. The trail led toward the sea, widening as it went. Soon it opened into a cleared circle some hundred feet in circumference. Involuntarily Mafatu started forward, then drew back with a sharp cry. What he beheld filled him with awe and it set him trembling.

He saw a series of wide stone terraces rising in a pyramid many feet high; on top of this pyramid a grotesque idol, hideously ugly, reared in the brilliant sunshine. It was an ancient idol, its contours softened with fungus and lichens, corroded by the rains of ages. The roots of convolvulus writhed about its base. No wind reached this hidden circle, and insects hummed in the hot air. Mafatu felt that he was stifling. His heart pounded. A *marae*—a Sacred Place. . . .

stopped, suddenly alert. They were splendid trees, three times his own height, with broad leaves that waved in the wind like tattered banners. But what caught and held his attention were the stems where fruit had grown: they had been severed cleanly by a knife! His heart gave a great thump and hung fire.

He examined the stems with care. Tree after tree had been stripped of fruit, and within a week, too. Who had cut these bananas? Well—he would have to find out! His lips were grim as he set about making his "sled." He bound the leaves together with a fibrous rope of vines and soon had a toboggan as long as his body. When Mafatu reached the lava slide, he set his sled in place and flung himself down upon it. With a shout and a shove he was off.

Down the natural slide he sped at terrifying speed. Trees whizzed past. Wind drove the air from his nostrils. The valley rushed to meet him. He brought up with a bump in a thicket of *cassi*. As he extricated himself from the thorns he was still breathing with excitement. *Aué*, but that was fun! He did not stop to think how he would regain his

swear it. My father, Tavana Nui, will be filled with pride at my homecoming. It is a vow that I take now, O Maui. I have spoken." . . .

The wind from the sea swept up around him, its voice warm and soft and reassuring in his ear. Maui, God of the Fishermen, had heard and answered.

Mafatu decided that before he retraced his steps to his own beach he would explore the opposite side of the island. The trail dropped from the plateau in a series of swift turns and spirals. The boy clambered down, slipping and sliding, catching hold of roots and vines to keep from falling. Far below, a cool dark stream wound its way through a sheltered valley. Perhaps in that valley he would find some people.

At the base of the ancient crater, long ridges of lava descended to the valley. Of a sudden Mafatu remembered an old tale out of his childhood: he had been told how the youths of Tahiti slid down the lava slide on sleds of giant leaves! The thought had scarcely struck him before he was wrenching off half a dozen great leaves from a near-by banana tree. Even as he did so he

look up to me and say: 'There goes Mafatu. He killed the wild boar single-handed!' And Tavana Nui, my father, will be filled with pride."

The prospect of returning home set another train of thought in motion. "I must find a tree, a tamanu, for my canoe," the boy said aloud. "I will burn it out, then make an adze of basalt to finish it. I'll plait a sail of pandanus. And oh, it will be a wonderful canoe!"

At that moment his eye fell upon a mango tree loaded with juicy fruit, and he plucked a fruit and sank his teeth into the rosy pulp. For a few seconds while he ate his fill and the juices ran down over his chin, he forgot all about his canoe; forgot that he needed shelter, food, fire, and weapons. Then, his hunger satisfied, his mind ran ahead again to the happy day when he would set sail for Hikueru, with all his demons put to rout and the bright flame of courage burning in his heart. Never again would he be called Mafatu, the Boy Who Was Afraid. He stood there taut with purpose, high above the demon sea.

"Maui, God of the Fishermen, hear me!" he pleaded. "I shall return home one day, I

what he would do! He was fired with excitement and set for the adventure. Then the thought of killing a wild boar in single-handed combat struck him dumb with wonder. Why, he would never have dreamed of such a thing in Hikueru! He was Mafatu, the Boy Who Was Afraid. . . . He set his jaw with fierce resolution. He had never known a man who killed a wild pig. But Grandfather Ruau, who had traveled as far as distant Tahiti, had told how the warriors of that island killed pigs in the mountains with naught but a knife for a weapon, bracing themselves so that the animal impaled itself upon the blade. It needed a strong arm, and a stouter heart. Could he, Mafatu, ever accomplish such a feat? Grandfather brought back with him a necklace made from the curling tusks of the boar, and Mafatu could remember to this day how the dark-blue scrolls of tattooing on the old man's copper flesh set off the handsome ivory teeth. That necklace was the envy of every man in Hikueru and its possession earned Grandfather much respect.

"I will make such a necklace for myself from the tusks," the boy promised bravely. "And when I return to Hikueru men will

hoarse cries as unceasing as the hum within a shell. Towering above Mafatu's head, the basalt cone of the island looked as soft in hue as an amethyst, broken and worn by a thousand years of wind and storm.

He observed that the barrier-reef encircled the entire island. There were only two openings in the reef through which canoes might enter. One opening lay on the side of the island where Mafatu had been cast ashore; the other was here to the southwest, facing the distant Smoking Island. In each case the opening was caused by a river which flowed from the mountain down into the lagoon; for the tiny coral polyp which builds up its ramparts from the floor of the sea cannot withstand fresh water. Wherever a river flows into the sea there will be a breach in the barrier-reef.

Mafatu jumped in alarm as a wild boar crashed through the undergrowth. It was so close that the boy caught a glimpse of its dark hide. Uri leaped in pursuit, barking furiously. Mafatu relaxed, and a smile crossed his face. Pig roasted underground in the hot oven stones of the *umu*— *Auê!* His mouth watered at the golden prospect. He would make a spear and kill the *puaa*, that's

He stood there looking off to the south-
west, and all at once his heart gave a jump,
and he strained forward. A cone-shape,
vague as a cloud upon the horizon, showed
him the existence of another high island. It
must have been fifty miles distant. As the
boy watched eagerly, scarcely daring to
believe the testimony of his eyes, he saw
what might have been a column of smoke
lifting high into the air from the peak of the
cone. He had heard Grandfather Ruau tell of
the Smoking Islands, the home of savage
tribes. They were the dark islands of the
eaters-of-men. Was that distant island one
such? Perhaps this very island upon which
he stood belonged to them, too! It was a
terrible thought.

As he stood there surveying his world, the
wind that swept up from the wide Pacific
beat hard against him, whistling in his ears.
Almost he had to lean against it to keep his
balance. It was a southwest wind that blew
straight from the Smoking Island, whipping
the sea to anger. Inside the barrier-reef the
water deepened and shoaled in changing
hues. Up here the whole world seemed con-
sumed with light and color. Far off a mist of
gulls drifted above the breaking surf, their

green gloom. Mafatu had never before seen woods like these, for Hikueru was open and wind-swept. These endless legions of trees seemed to close in upon him, imprison him with reaching arms, with heady odors, with eerie light and shadow. Ferns grew higher than a tall man's head; the roof of leaves was powdered with starry blossoms.

By the time Mafatu reached the plateau he was exhausted and his leg throbbed with pain. He lay down full length upon the volcanic rock and watched wild goats leaping from peak to peak high above his head and heard their shrill bleating in the clear air. When he caught his breath he sat up again and looked about. The plateau appeared to divide the island into halves. From his vantage point the boy could see its whole circumference. He was hoping desperately for some sign of human habitation, yet fearing it too; for who knew whether humans might prove friends or enemies? He almost hoped that the island were uninhabited, but if it were— He shivered as he realized his isolation. Even at sea in his small canoe he had not felt so utterly alone as he did here on this strange, high island. Everything about it was alien and forbidding.

he had to stop every now and then to catch his breath. Uri ran beside him, dashing off on this scent and that; the dog's shrill, sharp bark shattered the morning stillness.

For a quarter of a mile the coconuts held, beautiful trees that were more luxuriant than any in Hikueru. It was always thus in the rich soil of the volcanic islands. Then came a belt of breadfruit and wild bananas, of oranges and guavas and mangoes. The roots of the *mapé* trees—the island chestnut—twisted over the ground in strange, tormented shapes. Vines trailed like aerial ropes from the high branches where orchids bloomed, while little parakeets fled on swift wings and vanished in the

41

earth heaved with the sea's own motion, and he swayed to keep his balance. His leg still pained, and he would need the juice of limes to cauterize the coral wound, and *purau* leaves to make a healing bandage. Near by was a tree loaded with wild limes. He plucked half a dozen of the fruits, broke them on a bit of coral, and squeezed the juice into the wound. He winced as the caustic stung; but when he had bound on the leafy bandage with a twist of vine, it seemed that already his leg felt better. Soon the swelling would be gone.

Close at hand he discovered a rude trail made by wild pigs and goats in their wanderings across the mountain. The trail led up through the foothills to a high plateau which, the boy decided, would make a splendid lookout. From that point of vantage he would be able to survey his entire island and the sea for a distance of many miles.

He followed the path where it led back into the jungle, along the course of the swift-flowing stream. The trail sloped up sharply and Mafatu pulled himself up by roots and trailing lianas, now climbing, now crawling flat on his stomach. He found that

soft cooing of ghost-terns and the gentle plash of the cascade.

On his left hand, far offshore, the reef boomed to the charging surf; the curve of the beach reached out like two great arms to inclose the lagoon. Coconuts and pandanus trooped in shining legions to the very edge of the sea. A flight of green and purple parakeets flashed across the sky and vanished. There was no other sign of life. No voices of men; no laughter of children; no footprint in the sand.

The volcanic peak that formed the background of the island rose perhaps three thousand feet straight up out of the sea. It was the cone of a volcano long extinct. From its base ridges of congealed lava flowed down to the distant shore. Once, in the dim beginnings of the world, this mountain had belched forth fire and brimstone, spreading destruction over the land. But the forgiving jungle through fertile centuries had crept back up the slopes, clothing them in green, green.

The boy rose and stood stretching his stiff limbs. The water had restored him and he felt much stronger. But he found that the

tired body. Soon he must find food. . . . There were thousands of coconut trees on every hand, rich with green fruit; but Mafatu was not yet strong enough to climb. Later he would try it. Then he would make fire, too; and he must search the island to find out if it were inhabited; and there was a shelter to build. Oh, there was much to do! He hardly knew where to begin. But now it was enough just to lie there and feel strength returning, to know that he was safe from the sea. The sea. . . . He shuddered. Maui, God of the Fishermen, had carried him safely across the ocean currents.

"Uri!" the boy muttered thickly, "we're alive! It wasn't all a bad dream. It really happened."

The answering wag of his dog's tail was further assurance of reality. As Mafatu's brain cleared of cobwebs, a sudden thought brought him up swiftly: this silent island was not Tahiti. What island was it then? Did it . . . oh! did it belong to the black eaters-of-men? Were they even now watching him from secret places in the jungle, biding their time? He glanced about in apprehension. But the solitude was unbroken save for the

fell back upon one elbow. Uri lay close at hand, holding a robber-crab in his forepaws, cracking the tough shell and extracting the meat with gusto. There was Kivi, too, with his beak tucked back under his wing in sleep. Kivi, who had led his friends to this island. . . .

Mafatu pulled himself to a sitting position. The action called for more strength than he realized. He was giddy with thirst and there was an odd weakness in his limbs. His right leg was swollen and painful. He remembered then that he had banged it against the coral when the canoe struck. He discovered that there was a gash on his calf; he must take care of it, for coral wounds were poisonous.

The chuckle of the cascade reached his ears and made him aware of a stabbing need of water. He plunged his face into the pool and drank deeply. Then, prompted more by instinct than by conscious thought, he withdrew, to let the water run down his swollen throat sparingly, with caution. Its cool magic stole through his tissues, bringing with it new life and restoring force. He sighed, and sank back on the mossy bank, relishing the strength that quickened his

3
The Island

THERE WAS a fan of light spreading in the east. Mafatu stirred and opened his eyes. For a moment he lay there motionless in the cool mosses, forgetful of the events which had cast him up on this strange shore. Then it all came crowding back upon him, and he scarcely dared to believe that there was earth, solid earth beneath him; that once more Moana, the Sea God, had been cheated. He struggled to sit upright, then

rock a small cascade lost itself amid ferns and cool mosses. A ragged, strangling cry broke from Mafatu's throat. He stood upright to full height. Then he crashed to the mossy bank. One cheek lay in the cool water.

The moon lifted above the rim of the palms. It outlined in silver the form of a boy thin with hunger, naked as the daystar. It revealed a small wet dog, dragging himself across the beach to his master's side.

Mafatu lay without moving. Before Uri drank, he touched the boy's cheek with his hot muzzle.

flung over the barrier-reef, for here the water was scarcely troubled by wind or tide. Now he was swimming, swimming. . . . Somewhere ahead a strip of beach, salt-white in the darkness, lured him onward. His muscles did it of themselves. Only a will to live. A strip of sand white in the night. . . . He caught the gleam of a shark's belly, close at hand, but on he swam. His unhampered limbs moved freely through the water.

Of a sudden there was something solid beneath his feet. Sand. Mafatu stumbled, staggered, fell to his knees in the shallows. His lips moved in dry, soundless speech. Lying there with the water rippling and breaking over him, he pulled himself half upright, swayed onward. The palms, trooping to the edge of the beach, were motionless in the night air. All the world seemed to hold its breath as this boy climbed up out of the sea.

He fell to the sand once more, then, guided by he knew not what impulse, he dragged himself to the edge of the jungle. A murmur of water reached his ears, soft as a chuckle of pleasant laughter. Water, sweet water. . . . Down the face of an age-worn

34

quieting with sundown, like the reassuring hush of a mother to her child.

Night stole up from the face of the sea, enveloping the world. There was no moon, but the black sky was spangled with unguessed millions of stars: other worlds, other suns. To the watching boy, with land in the offing, they seemed closer and more friendly. The bottom star of the Southern Cross pointed to the end of the world. . . . A soft land breeze, heavy with a scent of flowers, waited out across the dark waters, tantalizing, bitter-sweet.

Weak with thirst, the boy drifted now into a merciful sleep. He struggled against it, like a weary swimmer fighting a rip-tide, but his head drooped and his eyes closed.

He was aroused at midnight by a thunderous tumult in his ears. Of a sudden he felt the canoe under him lifted and flung high into the air. Then down it crashed into splinters upon the reef. Boy and dog were hurled headlong into the boiling surf.

The shock of cold water brought Mafatu half back to consciousness. Blindly he struck out, fighting for survival. Uri— where was he? No sign of the dog. The boy was aware that the canoe must have been

blue-shadowed against the paler tone of the sky. Hour by hour, with every lift of the swells, the island rose higher and higher, filling Mafatu's soul with wonder. Hikueru, the only land he had ever seen, was as flat as his hand; but a great single peak crowned this strange island. Trees rose green and fair, tier upon tier of them, from the shoreline into the foothills of the purple mountain. Uri had caught the scent of the land now and was quivering with delight.

Then from far off came the first muffled thunder of the reef: the boom of the surf high-bursting on the barrier coral. That sound—was it the voice of Moana? "Someday, Mafatu, someday." . . . Involuntarily the boy shuddered. Would his ears never be free of the Sea God's threat?

Mafatu was powerless to guide his craft. He sensed that the current had quickened. He could only watch helplessly as the little canoe, swift as the following gulls, rushed to meet the tides of the island where they met and churned in a cross sea of conflict. Now across the swells came a sound like a chorus of ghostly fishermen weary with their day's toil: sea birds, always complaining, never at rest; while softer, yet rising above it was another sound—the voice of the reef itself,

higher, now revealed, now concealed it. That shadow on the horizon—it was land! The boy flung himself forward, shaking uncontrollably.

He seized Uri in his arms and lifted him up, laughing, crying: "Uri! Uri! It's land. *Land!*"

The dog sniffed at the air and a little whimper broke from him.

What island could this be? Was it Tahiti, the golden island, whose language was akin to that of Hikueru? Or was it, perhaps, one of the terrible dark islands of the eaters-of-men?

Now the current had a westward drift, and it was to the west that the dark islands lay. . . .

All day, as the canoe drifted onward, the boy watched the distant shadow-shape of land, not daring to take his eyes from it lest it vanish into the sea. Hunger and thirst were lulled into forgetfulness. There was only this one reality, land—escape from the sea. Weak as he was, he still clung to the thwart, his lips whispering a silent prayer of gratitude. With waning afternoon, the island took more distinct form as the canoe drifted nearer. It was high and peaked, its valleys

The brave thump of his dog's tail touched Mafatu profoundly. He caught the animal to him, while a new assurance, a new strength, flooded his being. If Uri could have courage to die, surely he, Mafatu, could not do less! In that instant he heard a whir and fury in the sky above, a beat of wide wings. . . . Looking upward, the boy's dulled eyes made out the widespread wings of an albatross, circling above the canoe.

"Kivi!" Mafatu cried hoarsely. "*Ai*, Kivi!"

Even as he spoke, the bird wheeled slowly, then flew off straight ahead for the distant horizon. The boy noticed then that the sea current was carrying him almost due southwest. Kivi's flight moved in exact parallel. Once more it seemed as if his albatross were leading him onward, just as he had led the canoe out of the passage of Hikueru.

Mafatu scanned the rim of the horizon; it looked as hard as the cut edge of a stone. But suddenly the boy's heart gave a great leap and he started forward. *It couldn't be!* It was a cloud. . . . But the sky was cloudless. Far off in the sea-shimmer lay something that was neither sea nor sky. The swells, lifting

Now the air was luminous with promise of another day. Out of the sultry mists the sea emerged, blue and violent. With the coming of this new day terror raised its head. Mafatu tried to fight it back, to deny its existence; but it gripped his heart with clammy fingers, tightened his throat. He flung himself flat on the floor of the canoe and buried his face in his arms. He must have cried out then. His voice was but a hoarse croak, yet it stirred Uri to life: the dog's ragged tail gave one feeble thump. With the ghost of a whimper the animal laid his hot nose against the boy's hand.

ing seas. Far away the wings of an albatross caught its gold as it wheeled and planed against the roof of heaven. The only hint of recent storm lay in the rough and tumbling waters. As the sun climbed through the hot hours of morning, it burned into the boy's body like the sacred fires of the great marae of Hikueru. Mafatu's skin blistered and cracked. His tongue swelled in his throat. He tried to call out a prayer to Maui, but his voice was thick; the sounds which came forth were no more than a hoarse cry. The canoe, stripped of sail and mast, without a paddle to guide it in the swift-racing current, twisted and shifted in the rushing waters.

As one hour merged into another there came moments of fitful, choking slumber, a growing agony of thirst for the boy and his dog. The sun burned into them like an inescapable eye. The current which held Mafatu's canoe fast in its grip was bearing it swiftly on toward its mysterious destination.

And thus the day passed, while night once more descended, bringing blessed release from the sun.

* * *

forward, wound his arms about the mid-thwart. It was the end of a world.

The wave passed. Stunned, gasping, Mafatu raised his head and looked about. For a second he could not believe that he still breathed and had being. He saw Uri wedged under the bow, choking for air. He pulled the dog free. Then he saw that his string of drinking nuts had vanished. His fish spear was gone. The knife that hung about his neck by a twist of bark had been torn away. Even his *pareu* of fiber tapa fell from his body as water soaked it through. He was naked, defenseless, without food or weapon, hurled forward on the breath of the hurricane. Numb of all feeling, empty as a shell, still he clung to life, and the hours droned by. . . .

So gradual was the storm's easing that at first the boy was unaware of it. The wind was blowing itself out, moving off into the empty spaces of the world. Uri crept toward the prostrate boy, quailing beside him, whimpering softly.

Night came and passed.

There was no morning mist to dim the splendor of the sunburst across the swing-

it had claimed his mother. How rightly he had feared it! Moana, the Sea God, had been biding his time. . . . "Someday, Mafatu, I will claim you!"

The boy lost all sense of time's passage. Every nerve became dulled by tumult. The wind howled above his head and still Mafatu clung to the lashed steering paddle; clung fast long after strength had vanished and only the will to live locked his strong fingers about the shaft. Even death would not loose the grip of those fingers. He held his little craft true to the wind.

There was a wave lifting before the canoe. Many the boy had seen, but this was a giant—a monster livid and hungry. Higher, higher it rose, until it seemed that it must scrape at the low-hanging clouds. Its crest heaved over with a vast sigh. The boy saw it coming. He tried to cry out. No sound issued from his throat. Suddenly the wave was upon him. Down it crashed. *Chaos!* Mafatu felt the paddle torn from his hands. Thunder in his ears. Water strangling him. Terror in his soul. The canoe slewed round into the trough. The boy flung himself

to the end of the boom in globes of fire that exploded and vanished, and in the awful moment of its being it revealed mountain shapes of dark water, heaving, shouldering. . . . How long could this frail craft of wood and sennit resist? Under the combined attack of wind and sea it seemed that something must inevitably give way. The wind shrilled a fiercer note. Spray stung the boy's flesh, blinded his eyes, chilled his marrow.

The sail went first—with a split and a roar. Fragments swept off on the back of the wind. The cords that held the mast hummed like plucked wires. Then with a rending groan the mast cracked. Before Mafatu could leap to cut it clear, it snapped off and disappeared in a churn of black water. The boy clung to the paddle, fighting to keep his canoe from turning broadside. Water swept aboard and out again. Only the buoyancy of tamanu kept the craft afloat. Uri cowered in the bow, half submerged, his howls drowned by the roar of the elements. Mafatu gripped his paddle for very life, an unreasoning terror powering his arms. This sea that he had always feared was rising to claim him, just as

paddle with hands that showed white at the knuckles. All around him now was a world of tumbling water, gray in the hollows, greenish on the slopes. The wind tore off the combing crests and flung the spray at the sky. Like advance scouts of an oncoming army, wind gusts moved down upon the canoe, struck at it savagely. So busy was Mafatu with the paddle that there was no time for thought. He called a prayer to Maui, God of the Fishermen:

"*Maui 'e! E matai tu!*"

Somehow the sound of his own voice reassured him. Uri lifted his head, cocked his ears, thumped his tail for a second. The canoe rose to the swells as lightly as a gull and coasted like a sled down the frothing slopes. What skill had wrought this small canoe! This dugout, hewn from the mighty tamanu tree. It swooped and yielded, bucked and scudded, one with the fierce element whose back it rode.

The sky darkened. A burst of lightning lit up the sea with supernatural brilliance. An instantaneous crack of thunder shattered the world. Lightning again, striking at the hissing water. Mafatu watched it with fascinated eyes. Now it was all about him. It ran

23

liquid trickle down his parched throat; more refreshing than spring water, cool on the hottest days, and as sustaining as food. The boy scooped out the gelatinous meat for Uri and the dog ate it gratefully.

The ocean current which held the canoe in its grip seemed to have quickened. There was a wind rising, too, in little puffs and gusts. Now the canoe heeled over under the sudden attack, while Mafatu scrambled onto the outrigger to lend his weight for ballast; then the wind dropped as suddenly as it appeared, while the canoe righted itself and the boy breathed freely once again. He searched the skies for Kivi. His albatross might have been one of a thousand sea birds flying against the roof of the sky, or he might have vanished utterly, leaving his friends here in solitary space. The bird had led Mafatu out through the reef-passage at Hikueru into the open ocean, and now, it seemed, had deserted him.

A storm was making, moving in out of those mysterious belts which lie north and south of the equator, the home of hurricanes. The wind shifted a point, bringing with it a heavy squall. Mafatu lowered the sail on the run and gripped the steering

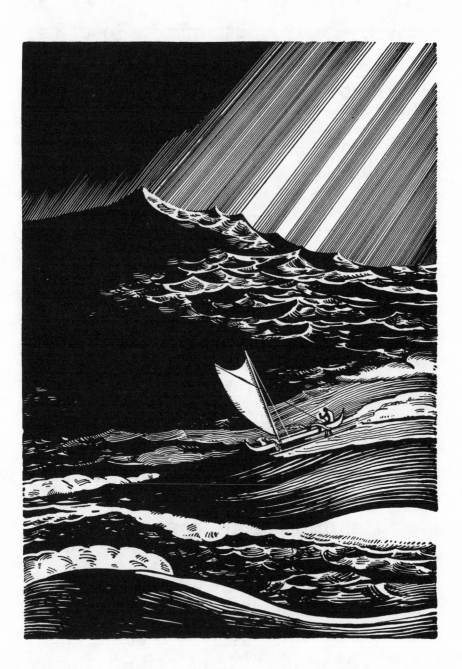

his father's heart? Would he believe that Moana, the Sea God, had claimed his son at last?

It was an ominous, oppressive world at this season of storm. Half a mile distant a whale heaved its varnished hulk to the surface, to throw a jet of vapory mist high into the air; then it submerged, leaving scarcely a ripple to mark its passage. A shoal of flying fishes broke water, skimming away in a silver shimmer of flight. A dolphin sped after them, smooth-rolling in pursuit, so close that the boy could hear the sound of its breathing. This world of the sea was ruled by Nature's harsh law of survival. Mafatu knew the sea with an intimacy given to few. He had seen fleets of giant mantas whipping the lagoon of Hikueru to a boiling fury; he had seen the mighty cachalot set upon by killer-whales and torn to ribbons almost in the blink of an eye; once he had seen an octopus as large as the trunk of a tamanu, with tentacles thirty feet long, rise from the mile-deep water beyond the barrier-reef. . . . *Ai*, this sea!

Mafatu opened one of the green drinking nuts and tilted back his head to let the cool

through the length and breadth of the Pacific: the *Ara Moana*, Paths of the Sea, as the Ancients called them. They were the ocean currents that had carried the Polynesian navigators from island to island in the childhood of the world. Mafatu was drifting farther and farther away from his homeland.

With wide-flapping wings Kivi rose from the bow of the canoe. In ascending spirals the bird climbed higher and higher, until at last he was no more than a gray speck against the lighter gray of the sky. Mafatu watched his albatross disappear and felt a desolation flood his heart. Now there was only Uri to keep him company in this hostile world of sky and sea. Uri. . . . The yellow dog lay curled up in the shadow of the bow, opening one eye from time to time to look at his master. Wherever Mafatu went, Uri, too, would go.

All around, as far as the eye could reach, were wastes of leaden water. The canoe was the moving center of a limitless circle of sea. The boy shuddered. His fingers gripped the paddle convulsively. He thought of Kana and the other boys—what would they say when they learned that he had disappeared? And Tavana Nui—would there be sorrow in

19

2
The Sea

DAY BROKE over a gray and dismal world.
The canoe lifted and fell idly on the glassy
swells. Mafatu looked back over his shoul-
der, searching the horizon for a last glimpse
of Hikueru; but the atoll had vanished, as if
to hide itself forever from his concern.

The matting sail slatted uselessly. But
there seemed to be no need of a sail: the little
canoe was riding one of the mysterious
ocean currents that flow in their courses

its churn, swept it forward like a chip on a millrace. No turning back now. . . .

The boy was aware of a sudden whir and fury in the sky above, a beat of mighty wings. Startled, he glanced upward. There was Kivi, his albatross. Mafatu's heart lifted. The bird circled slowly in the moonlight, its wings edged with silver. It hovered for a moment just over the bow of the canoe, then it rose easily, lightly in its effortless flight. Out through the passage in the reef. Out into the open ocean.

Mafatu gripped the steering paddle and followed.

was only Kivi—Mafatu would miss his albatross. He scanned the dark sky for sight of the bird, then gave it up and turned away.

The lagoon was as untroubled as a mirror. Upon its black face the stars lay tracks of fire. The boy shoved off and climbed into the stern. Noiselessly he propelled the canoe forward, sending it half a length ahead with each thrust of his paddle. As he drew nearer to the barrier-reef, the thunder of the surf increased. The old, familiar dread of it struck at his stomach's pit, and made him falter in his paddling. The voices of the Old Ones were fainter and fainter now.

The reef thunder mounted: a long-drawn, hushed yet mighty sound that seemed to have its being not in the air above but in the very sea beneath. Out beyond lurked a terrifying world of water and wind. Out there lay everything most to be feared. The boy's hands tightened on his paddle. Behind him lay safety, security from the sea. What matter if they jeered? For a second he almost turned back. Then he heard Kana's voice once more saying: "Mafatu is a coward."

The canoe entered the race formed by the ebbing tide. It caught up the small craft in

would hear his father say: "Here is my son Stout Heart. A brave name for a brave boy." . . . Standing there with clenched fists, Mafatu knew a smarting on his eyelids and shut his eyes tight, and sank his teeth into his lower lip.

Far off in the *himene* house the Old Ones were singing. Their voices filled the night with rich sound. They sang of long voyages in open canoes, of hunger and thirst and battle. They sang the deeds of heroes. The hair on the boy's damp forehead stirred; the long-drawn mutter of the reef sounded its note of warning in his ears. At his side, Uri touched his master's hand with a cold nose. Mafatu pulled the dog close.

"We're going away, Uri," he whispered fiercely. "Off to the south there are other islands." . . .

The outrigger canoes lay drawn up on the beach like long slim fish. Silent as a shadow, the boy crossed the sand. His heart was hammering in his throat. Into the nearest canoe he flung half a dozen green drinking nuts, his fish spear. He gave his *pareu* a brave hitch. Then he picked up a paddle and called to Uri. The dog leaped into the bow. There

have tried to be friendly with him. But he is good only for making spears. Mafatu is a coward."

The boys disappeared down the moonlit beach. Their laughter floated back on the night air. Mafatu stood quite still. Kana had spoken; he had voiced, once for all, the feeling of the tribe. Mafatu—Stout Heart—was a coward. He was the Boy Who Was Afraid.

His hands were damp and cold. His nails dug into his palms. Suddenly a fierce resentment stormed through him. He knew in that instant what he must do: he must prove his courage to himself, and to the others, or he could no longer live in their midst. He must face Moana, the Sea God—face him and conquer him. He must.

The boy stood there taut as a drawn arrow awaiting its release. Off to the south somewhere there were other islands. . . . He drew a deep breath. If he could win his way to a distant island, he could make a place for himself among strangers. And he would never return to Hikueru until he should have proven himself! He would come back with his head high-held in pride, and he

shadow of a pandanus. A group of boys were pulling their canoes above high watermark, and laying their plans for the morrow. Their voices were shrill with eagerness.

"Tomorrow at daybreak . . ." one was saying.

"There'll be Timi and Tapu and Viri . . ."

"*Auê!*" another voice broke in. "It's work for us all. How else will we become fishermen and warriors? How else will we feed our families and keep the tribe alive?"

"True! Hikueru is too poor. There are only the fish from the sea. A man must be fearless to provide food. We will all go— every one of us!"

Mafatu, standing tense in the shadows, heard a scornful laugh. His heart contracted. "Not all of us will go," he heard Kana scoff. "Not Mafatu!"

"Ha! He is afraid."

"He makes good spears," offered Viri generously.

"Ho! That is woman's work. Mafatu is afraid of the sea. *He* will never be a warrior." Kana laughed again, and the scorn of his voice was like a spear thrust through Mafatu's heart. "*Aiá!*" Kana was saying. "I

13

"Wait, Kana! I'll go! I'll try—" But the words would not come. Kana had gone. Tomorrow he and all the other boys would be taking their canoes out beyond the reef. They would return at sunset, loaded down with bonitos, their faces happy, their shouts filling the dusk. Their fathers would say: "See what a fine fisherman is my son! He will be a Chief one of these days." Only Tavana Nui would be silent. *His* son had not gone.

That night a new moon rose above the edge of the sea, silvering the land with a bloom of magic. Wandering along the outer beach with Uri, Mafatu heard laughing voices and drew hastily into the black

swordfishes and tiger-sharks, progressing from one stage to a higher. Every boy in the village sharpened his spear, tested the shaft, honed his shark knife. Every boy, that is, except Mafatu.

Kana stopped one afternoon to watch Mafatu at work on his nets. Of all the youths of his own age, Kana alone had been friendly. Sometimes he even stayed behind when the others were fishing to help the boy with his work.

"The bonitos have begun to run, Mafatu," Kana said quietly.

"Yes," the other returned, then fell silent. His fingers faltered as they flew among the sennit fibers of the net he was making.

"My father brought back word from the reef today," Kana went on. "Already there are many bonitos out there. Tomorrow we boys will go after them. That's our job. It will be fun, eh?"

Mafatu's knuckles whitened. His ears pounded with the swift fury of the sea. . . .

"That will be fun, won't it?" Kana insisted, watching Mafatu closely. But the boy made no answer. Kana started to speak; he stopped, turned impatiently, and walked away. Mafatu wanted to cry out after him:

11

other. Perhaps because it was different from its kind, the older birds were heckling and pestering the fledgling. Something about that small bird trying to fight off its more powerful fellows touched the boy's heart. He picked it up and carried it home—caught fish for it in the shallows of the lagoon. The bird followed Mafatu and Uri about, limping on its one good leg. At length, when the young albatross learned to fly, it began to find its own food. In the air it achieved perfection, floating serenely against the sky while Mafatu followed its effortless flight with envious eyes. If only he, too, could escape to some world far removed from Hikueru!

Now, once more, it was the beginning of the season of storms. Men scanned the skies anxiously, watching for the dreaded signs which might spell the destruction of their world. Soon the great bonitos would be swimming beyond the reef—hundreds, thousands of them—for they came each year at this time with the unfailing regularity of the tides. They were held to be the special property of young boys, since it was by killing them that a youth learned to kill the

setting his scalp atingle: pictures of himself, a babe, clinging to his mother's back . . . sharks cruising. . . . And so overcome would he be at the remembrance of that time that he would drop his spear overboard, or let the line go slack at the wrong moment and lose the fish.

It was obvious to everyone that Mafatu was useless upon the sea. He would never earn his proper place in the tribe. Stout Heart—how bitter the name must taste upon his father's lips!

So, finally, he was not allowed to fare forth with the fishermen. He brought ill luck. He had to stay at home making spears and nets, twisting coir—the husk of the coconut— into stout sharkline for other boys to use. He became very skillful at these pursuits, but he hated them. His heart was like a stone in his breast.

A nondescript yellow dog named Uri was Mafatu's inseparable companion—Uri with his thin coat which showed his ribs, and his eyes so puzzled and true. He followed the boy wherever he went. Their only other friend was Kivi, an albatross. The boy had once found the bird on his lonely wanderings. One of Kivi's feet was smaller than the

the men in battle against warriors of other islands? Mafatu's father heard the whispers, and the man grew silent and grim.

The older people were not unkind to the boy, for they believed that it was all the fault of the *tupapau*—the ghost-spirit which possesses every child at birth. But the girls laughed at him, and the boys failed to include him in their games. And the voice of the reef seemed pitched for his ears alone; it seemed to say: "You cheated me once, Mafatu, but someday, someday I will claim you!"

Mafatu's stepmother knew small sympathy for him, and his stepbrothers treated him with open scorn.

"Listen," they would mock. "Moana, the Sea God, thunders on the reef. He is angry with us all because Mafatu is afraid!"

The boy learned to turn these jibes aside, but his father's silence shamed him. He tried with all his might to overcome his terror of the sea. Sometimes, steeling himself against it, he went with Tavana Nui and his stepbrothers out beyond the reef to fish. Out there, where the glassy swells of the ocean lifted and dropped the small canoe, pictures crowded into the boy's mind,

hungry fingers. The little boy was too weak even to cry. At hand lay a cracked coconut; the woman managed to press the cool, sustaining meat to her child's lips before she died.

Sometimes now, in the hush of night, when the moon was full and its light lay in silver bands across the pandanus mats, and all the village was sleeping, Mafatu awoke and sat upright. The sea muttered its eternal threat to the reef. The sea. . . . And a terrible trembling seized the boy's limbs, while a cold sweat broke out on his forehead. Mafatu seemed to see again the faces of the fishermen who had found the dead mother and her whimpering child. These pictures still colored his dreams. And so it was that he shuddered when the mighty seas, gathering far out, hurled themselves at the barrier-reef of Hikueru and the whole island quivered under the assault.

Perhaps that was the beginning of it. Mafatu, the boy who had been christened Stout Heart by his proud father, was afraid of the sea. What manner of fisherman would he grow up to be? How would he ever lead

hissing with spray. The poles of the out-
rigger were torn from their thwarts. The
woman sprang forward to seize her child as
the canoe capsized. The little boy gasped
when the cold water struck him. He clung to
his mother's neck. Moana, the Sea God, was
reaching up for them, seeking to draw them
down to his dark heart. . . .

Off the tip of Hikueru, the uninhabited
islet of Tekoto lay shrouded in darkness. It
was scarcely more than a ledge of coral,
almost awash. The swift current bore
directly down upon the islet.

Dawn found the woman still clinging to
the *purau* pole and the little boy with his
arms locked about his mother's neck. The
grim light revealed sharks circling, cir-
cling. . . . Little Mafatu buried his head
against his mother's cold neck. He was filled
with terror. He even forgot the thirst that
burned his throat. But the palms of Tekoto
beckoned with their promise of life, and the
woman fought on.

When at last they were cast up on the
pinnacle of coral, Mafatu's mother crawled
ashore with scarcely enough strength left to
pull her child beyond reach of the sea's

4

mother had taken him out to the barrier-reef to search for sea urchins in the reef pools. There were other canoes scattered at wide intervals along the reef. With late afternoon the other fishermen began to turn back. They shouted warnings to Mafatu's mother. It was the season of hurricane and the people of Hikueru were nervous and ill at ease, charged, it seemed, with an almost animal awareness of impending storm.

But when at last Mafatu's mother turned back toward shore, a swift current had set in around the shoulder of the reef-passage: a meeting of tides that swept like a millrace out into the open sea. It seized the frail craft in its swift race. Despite all the woman's skill, the canoe was carried on the crest of the churning tide, through the reef-passage, into the outer ocean.

Mafatu would never forget the sound of his mother's despairing cry. He didn't know then what it meant; but he felt that something was terribly wrong, and he set up a loud wailing. Night closed down upon them, swift as a frigate's wing, darkening the known world. The wind of the open ocean rushed in at them, screaming. Waves lifted and struck at one another, their crests

3

them across the Pacific in their sailing canoes, before the dawn of recorded history, not knowing where they were going nor caring what their fate might be, still sang its song of danger in their blood. There was only courage. A man who was afraid—what place had he in their midst? And the boy Mafatu—son of Tavana Nui, the Great Chief of Hikueru—always had been afraid. So the people drove him forth. Not by violence, but by indifference.

Mafatu went out alone to face the thing he feared the most. And the people of Hikueru still sing his story in their chants and tell it over the evening fires.

It was the sea that Mafatu feared. He had been surrounded by it ever since he was born. The thunder of it filled his ears; the crash of it upon the reef, the mutter of it at sunset, the threat and fury of its storms—on every hand, wherever he turned—the sea.

He could not remember when the fear of it first had taken hold of him. Perhaps it was during the great hurricane which swept Hikueru when he was a child of three. Even now, twelve years later, Mafatu could remember that terrible morning. His

1
Flight

IT HAPPENED many years ago, before the traders and missionaries first came into the South Seas, while the Polynesians were still great in numbers and fierce of heart. But even today the people of Hikueru sing this story in their chants and tell it over the evening fires. It is the story of Mafatu, the Boy Who Was Afraid.

They worshiped courage, those early Polynesians. The spirit which had urged

CONTENTS

First published 1940 in the United States
by Macmillan Publishing Company.

Published in large print 1989 by arrangement
with Macmillan Publishing Company,
a division of Macmillan, Inc.

Library of Congress Cataloging-in-Publication Data

Sperry, Armstrong, 1897-
 Call it courage.
 Summary: Relates how Mafatu, a young Polynesian boy whose name means Stout Heart, overcomes his terrible fear of the sea and proves his courage to himself and his people.
 1. Large type books. [1. Folklore—Polynesia. 2. Courage—Folklore. 3. Survival—Folklore. 4. Large type books.] I. Title.
PZ8.1.S75Cal 1989 398.22'0996 89-23862
ISBN 1-55736-147-9

10 9 8 7 6 5 4 3 2 1

Large print edition design by Terri Wright.
Composed in 18/20 pt. Plantin by the TypeStudio,
Santa Barbara, California. Printing by
Braun-Brumfield, Inc., Ann Arbor, Michigan.

This book is Smyth-sewn and printed on acid-free paper ∞ .
Manufactured in the United States of America.

Call It Courage

Armstrong Sperry

Illustrations by the Author

CORNERSTONE BOOKS

ABC-CLIO

Santa Barbara, California
Oxford, England

CALL IT COURAGE